Environmental Management

Basharat Mushtaq · Suhaib A. Bandh ·
Sana Shafi

Environmental Management

Environmental Issues, Awareness and Abatement

Basharat Mushtaq
Department of Environment and Water
Management
Sri Pratap College Campus,
Cluster University
Srinagar, Jammu and Kashmir, India

Suhaib A. Bandh
Post Graduate Department of Environmental
Science
Sri Pratap College Campus, Cluster University
Srinagar, Jammu and Kashmir, India

Sana Shafi
Department of Higher Education
Sri Pratap College, Cluster University
Srinagar, Jammu and Kashmir, India

ISBN 978-981-15-3815-5   ISBN 978-981-15-3813-1   (eBook)
https://doi.org/10.1007/978-981-15-3813-1

This Springer imprint is published by the registered company Springer Nature Singapore Pte Ltd.
The registered company address is: 152 Beach Road, #21-01/04 Gateway East, Singapore 189721, Singapore

Contents

About the Authors

Basharat Mushtaq graduated from Sri Pratap College Srinagar and received his master's degree in environmental science from the University of Kashmir, Srinagar. Specialized in aquatic ecology, he subsequently completed his doctorate in environmental science at Barkatullah University, Bhopal, Madhya Pradesh. Dr. Basharat has worked as a research associate in Environmental Impact Assessment/Environmental Management projects sponsored by various national and international agencies, e.g., the NHPC, WAPCOS, NRSC-ISRO, SDA, PDA, GDA, and YDA. He has also worked as a Senior Research Fellow in a major research project on Lake Ecology funded by LAWDA, J & K Government, Srinagar. He has published many research papers and book chapters in prominent national and international journals. Dr. Basharat has worked as a faculty member at the Cluster University Srinagar and as a Guest Faculty member at the Central University of Kashmir. Presently, he is working as a Lecturer in Environmental Science at Govt. Sri Pratap College Campus, Cluster University Srinagar.

Suhaib A. Bandh is an Assistant Professor at Govt. Sri Pratap College Campus, Cluster University Srinagar, where he teaches environmental science in graduate and integrated courses. A member of the Academy of Plant Sciences, India, and the National Environmental Science Academy India, he has participated in a number of national and international conferences held by prominent scientific bodies in India and abroad.

Dr. Bandh, a doctorate in Environmental Science has a number of scientific publications to his credit which attest to his scientific insight, fine experimental skills and outstanding writing skills. Dr. Bandh has already completed a book on Freshwater Microbiology with Elsevier Press and Perspectives of Environmental Science with Callisto References.

Sana Shafi is an Assistant Professor at Sri Pratap College, Srinagar, teaching environmental science to postgraduate and integrated courses. She holds a master's degree in environmental science and an M. Phil and Ph.D. degree in the same subject from the University of Kashmir, Srinagar. She has participated in a number of national and international conferences held by different reputed scientific bodies. She has a number of scientific publications to her credit. She has published her research findings in some highly reputed and high-impact journals, which attest to her scientific insight, fine experimental skills, and outstanding writing skills.

List of Figures

List of Tables

Management of Water Resources

1

Abstract

Water resources which are potentially useful to humans and the rest of the world, are not only important for the existence of life on earth but also for ensuring sustainable development of the world, as their effective and sustainable management is effectively linked to various sectors of our economy including the industrial, agricultural, power, fisheries, recreational and environmental, transportation, domestic and household sectors. The resources and the services provided by them have benefited both people and their economies for many centuries, yet in many countries people lack the basic drinking water and sanitation needs. Therefore, management of water resources has become a critically important issue for their future use, for their protection from overexploitation and pollution and for the prevention of disputes over water bodies. It is the proper planning, development, and management of water resources both in terms of quality and quantity across all water uses which include the infrastructure, policies, protocols, incentives, information, and institutions for its proper guidance and support.

Keywords

Water resources · Water resource management · Watershed management · Rainwater harvesting · Water resource distribution

Water being the most valuable and renewable natural resource is in a great trouble these days because the demand of this resource is inclining continuously while its supply is declining. Like air, it is an essential constituent of the life-support system as all living beings depend on water for their existence. Due to increased industrialization, modernization, urbanization, and other such phenomena the demand for water has increased manifold in the cities which has further increased due to the increased demand of water for removing different kinds of wastes and sewerages. If we look at India's water resources from a global context, India possesses only 4% of the water resources whereas it houses 16% of the world's human population indicating that the per capita water availability in India is quite low, although India is blessed with a good annual rainfall of about 200 cm (Anonymous 1999). Here the quantum of available water is about 6 trillion m^3 which is among the largest for a country of comparable size in the world. But due to wastage and inefficient water management systems, along with unequal distribution of rainfall, a large part of the country still suffers water scarcity during the dry months (Khullar 1999) of the year.

On planet earth, water is available in the oceans, in the atmosphere, in and on the land and fractured rocks of the earth's crust. The movement of water is driven by the solar energy from one location to another, as it circulates between the lithosphere and the atmosphere through the processes of evaporation and precipitation (Linsley and Franzini 1979). The oceanic system is the largest reservoir of water, but since it is saline it is not readily usable for the requirements of human survival. The freshwater portion is just a fraction of the total water available and its distribution is highly uneven with most of it locked in the frozen polar ice caps.

Unique Features of Water
The unique features that make water a unique resource are:

(a) It exists as a liquid over a wide range of temperature (0–100 °C).
(b) Due to its highest specific heat, it cools down and warms up slowly without causing temperature shocks to the aquatic life.
(c) Due to its ability to change the state of its existence it can readily be found in the gaseous, liquid, or solid form on the planet earth.
(d) Due to its high latent heat of vaporization it takes a huge amount of energy to get vaporized and produces a cooling effect on evaporation.
(e) It acts as an excellent solvent and hence acts as a good carrier of nutrients, including oxygen.
(f) It easily rises even to the tallest possible trees due to high surface tension and cohesion.

Sources of Water

Precipitation that most commonly occurs in the form of rain and snow acts as a major source for both the groundwater stored under earth's surface in fractured rocks and aquifers as well as the surface water stored in ponds, lakes, and streams. However, its distribution is quite varied with many locations having a little of it while some others having a plenty of it. Corresponding to the annual precipitation including the 4000 BCM (billion cubic meters) of snowfall, the average annual rainfall of India is around 1170 mm. Out of this amount only 1869 BCM flows in rivers and only 1123 BCM (433 BCM from groundwater and 690 BCM from surface water) is assessed as the average annual utilizable water due to certain constraints. The total water use which once was only 634 BCM (of which 83% was used for irrigation purposes) was projected to rise to 813 BCM by 2010, 1093 BCM by 2025, and 1447 BCM by 2050, against a utilizable quantum of 1123 BCM. Thus it can be expected that the demand of water will outstrip its availability in another 30–40 years. The water reserves like polar ice caps, oceans, rivers, lakes, streams, and ponds contain 324 million cubic miles of water while some 2 million cubic miles of water lie below in the groundwater deposits and there are another 3100 cubic miles of water present in the atmosphere as water vapors (Linsley and Franzini 1979).

Water in solid form (held in polar ice caps and glaciers) constitutes 2.14% of total global water resource and the Himalayas in the Indian subcontinent is a major contributor to this. Twelve major rivers of the Ganga, Indus, and Brahmaputra river systems also originate from the permanent snow cover while other 44 medium and 55 minor fast flowing river systems are mainly fed by the monsoons (Howard and Tchobanogloss 2002).

Major distribution of water resources:

- Surface water bodies contain *324 million cubic miles* of water.
- Underground water reserves contain *2 million cubic miles* of water.
- Atmosphere contains *3100 cubic miles* of water as water vapors.

Classification of Water Resources

Water resources can be classified into three main groups (Fig. 1.1):

1. Surface water resources
2. Groundwater resources
3. Water in dried form

Surface Water

The freely available water found on the surface of earth in the form of rivers, rivulets, streams, ponds, and lakes is known as surface water. With precipitation as its main source, melt water also contributes sizeable amount of water to the surface water which is largely used for industrial, irrigational, agricultural, navigational, and public water supply purposes. It can be divided into two types:

- Lotic (running) water
- Lentic (standing) water

Lotic Waters

The term lotic which has been derived from a word "lavo" meaning "to wash" represents the running water bodies including streams, rivers, runoff, and spring. It is

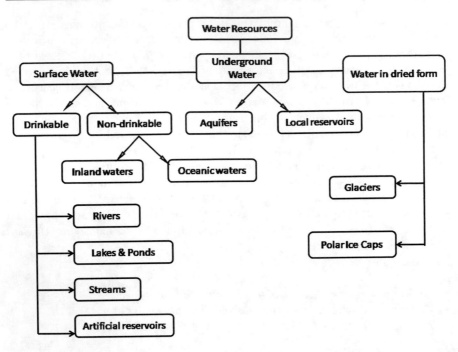

Fig. 1.1 Classification of water resources

that water where an entire water body moves in a definite direction (Marsh and Fairbridge 1999). As the water in such systems is in a state of constant motion, the basic function of lotic water systems is assumed to carry the excess rainwater back to the oceans. Lotic ecosystems are important in a much diverse and often complex way as they process energy, organic matter, nutrients, and other chemicals; receive materials from the land and air, transform them, and/or transport them to the oceans or to closed inland basins; and provide sustenance, protection, and corridors for movement of terrestrial plants and animals, including humans. The strength of water flow that varies between the different systems ranging from the slow backwaters to the torrential rapids is a key factor influencing the overall ecology of these systems. It further varies within the systems and is also subjected to chaotic turbulences which result in the divergence of flow. The mean flow rate in a lotic water body varies on the basis of obstructions, sinuosity, incline gradient, and the variability of friction with the sides and bottom of the channel which is further affected by the snowmelt, direct precipitation, and groundwater. Through depositions and erosion the lotic systems can change the shape of the streambeds and can create a variety of habitats including glides, pools, and riffles.

Rivers

Rivers, whose development is the work of ages, are formed along more or less defined channels to drain all the water received in the form of rainfall and melting of snow from the land. Most of the rivers originate from glaciers and therefore, at the headwaters, the rivers are usually cold and full of oxygen. Along the path down the mountains, rivers become broader, warmer, slower, and wider and contain a lesser content of oxygen. Further, a river changes with the change in the properties of land and climate through which it runs.

Streams

Rainwater infiltrates into the soil and subsequently joins the groundwater storage. When the natural relief is such that the ground surface at any point falls below the top surface of groundwater reservoir there exists greater hydrostatic pressure in the soil mass and the groundwater under pressure finds its way through the soil into the depression forming a stream. Streams may be divided into torrent streams and intermittent streams. Torrent streams are those streams which carry only the surface runoff and hence water flows through them only during the period of rainfall and the subsequent period of runoff. However, in the intermittent streams flow of water occurs only in rainy seasons and they dry out when the rainy season is over. It is so because frequent rains temporarily raise the water table during rainy seasons.

Lentic Water

Lentic waters are those closed systems which contain stagnant water. They are usually formed in small or large depressions on earth's surface which didn't possess any exit for the water to flow out. Lentic water entails a body of standing water which includes a variety of water bodies ranging from small shallow temporary pools, ponds, lakes, and marshes to enormous impoundments created for water storage (Marsh and Fairbridge 1999). Lentic waters usually decay, decompose, or persist as such within the lentic water body and the natural process like succession in the long run changes such a lentic water body into a marsh, a swamp, a wetland, and finally into a dry land. Some lentic ecosystems contain water with a higher content of salts (like the Pangong Lake in Jammu and Kashmir and the Great Salt Lake in Utah) while some contain the water which is fresh. Lentic ecosystems too display the vertical stratification based on the temperature and the amount of light penetration.

Lakes

Lakes are cavities in the soil, collecting waters conveyed to them by rivers from the catchment area or by atmospheric precipitations and groundwater. These cavities may have been produced in the earth's crust due to the occurrence of some catastrophic events in distant geological era (tectonic and volcanic lakes) or in relatively recent times (landslide lakes) or else they may have originated in the most recent geological era by some slow changes taking place in the earth's crust (glacial lakes, plain lakes, coastal lakes, karstic lakes, etc.).

Lake basins originate through a wide variety of natural and anthropogenic processes with some of them being cataclysmic (like volcanism) and some of them being gradual and imperceptible (like tectonic movements) while others being rare and extraordinary (like meteoritic impact) (Likens 2009).

A continuous supply water, an environmental force, and the subjective reshaping of the terrain into a closed basin (depression) are a few common factors responsible for the origin of lakes and it is due to the frequent merger of these three elements that millions of natural lakes are existent today on the landscape. To us most of the lakes on the landscape are a permanent feature; however, by contrast, on a geological time scale, most of the lakes are fleeting and hence ephemeral. According to the frequency of overturn the lakes (Hutchinson 1957) may be classified as follows:

- *Monomictic*: Lakes in which the overturn occurs once a year are known as monomictic lakes. These are temperate lakes which do not freeze.
- *Dimictic*: Lakes in which the overturn occurs twice a year are known as dimictic lakes. These are again the temperate lakes which do not freeze.
- *Polymictic*: These lakes are holomictic and have shallow waters. Their lesser depth prevents the development of thermal stratification and hence their waters mix regardless of the season.
- *Amictic*: These lakes have water whose surface is covered with ice throughout the year which prevents the mixing of water beneath, and therefore allows such lakes to exhibit inverse cold water stratification where water temperature increases with increasing depth.
- *Oligomictic*: These are the deep and tropical lakes which show poor mixing.
- *Meromictic*: These lakes have layers of water that do not intermix. Such types of lakes are mainly oligomictic but sometimes they are also deep monomictic and dimictic in nature.

Reservoirs

An artificial lake in which water is stored for irrigational, industrial, domestic, power generation, flood control, and other purposes is known as a reservoir (Merriam–Webster dictionary). A reservoir or artificial storage is found when some obstruction like a dam is constructed at the narrowest point of a valley to store large quantity of water behind and the water stored can be very conveniently used for various purposes with provisions of suitable hydraulic structures. Besides storing water, the reservoirs function to raise the level of water to be diverted into a pipeline or a canal or to increase the hydraulic head which is an expression of the water pressure (Likens 2009). The volume of a reservoir can usually be defined by its live storage capacity and dead storage capacity. The entire volume of water that can be withdrawn from the reservoir is known as live storage capacity while the volume of water that remains in the reservoir when it is emptied to its low water level is known as dead storage capacity.

Groundwater

Nearly 70% of the rainwater flows into rivers, lakes, drains, ponds, etc. and 10% evaporates. The remaining 20% soaks into soil and reaches certain distance below the ground through soil particles, and forms groundwater. So, the rainfall becomes the main source of groundwater that is also called as plutonic water. Beyond certain levels water cannot penetrate because of saturation of water table, i.e., the depth below the ground where all the soil particles are filled with water molecules (Shah 2005).

Groundwater contributes a significant amount of usable water in those areas where surface water is scarce. However, its availability depends upon the amount and nature of rainfall along with the nature and slope of land. In the areas of high rainfall and porous rocks, easy percolation makes underground water easily available in huge quantities even at shallower depths.

Origin of Groundwater

Total water existing on earth is 13,84,12,0000 km^3, out of which 8,00,0042 km^3 occur in the form of groundwater and 61,234 km^3 occur in the form of soil moisture (Bhattacharyya et al. 2015), which together constitute the subsurface quantity of water. Groundwater, that mainly comes from the following three sources is stored in different layers of earth that infiltrates through the pores and fissures of the permeable rocks.

1. *Meteoric water*: It is the main source of groundwater that is received in the form of rains and snow. This water infiltrates from the surface through fissures, pores, and joints of rocks till it reaches the non-permeable rocks and forms the groundwater.
2. *Connate water*: Connate water also called as sedimentary water exists in pores and cavities of sedimentary rocks of seas and lakes.
3. *Magmatic water*: Water that occurs due to the condensation of vapor as a result of volcanic action at the time of entering hot rocks is known as magmatic water.

Open Wells

An open well is a lined or unlined hole in the ground that accesses the shallowest groundwater available in the local area. Wells typically get water from the "unconfined shallow aquifers", which are the non-pressurized water bearing soil or rock layers at shallow depths.

Open wells are the earliest tool invented by mankind to access groundwater. Historically, they have been used primarily for irrigation, domestic, and non-domestic purposes. Over centuries India's open wells have played a major role in conservation and optimum usage of water during times of water shortage.

Tube Wells

A well in which a long and wide stainless steel or plastic pipe is bored into an aquifer is known as a tube well. The required depth of a tube well depends on the level of the water table (Mays 2001). In tube wells the metal pipe driven in ground is perforated to allow only clear water to enter the hole and as an alternative, a wire net is wrapped on the cylindrical frame. The best and most commonly adopted practice is to provide a pipe with fairly big perforations and surrounding that in a wire net or a strainer with smaller openings whose finer openings exclude the objectionable soil particles from entering the tube well. So as to maintain a constant inflow velocity, the total area of the openings in the metal tube and the strainer is kept the same.

Artesian Wells

A well that does not require a pump-like structure to bring out the water to the surface is known as an artesian well. Its water flows to the surface under the influence of the natural pressure of the underlying aquifers which forces water to the surface without any mechanical assistance (USDIBR 1977). Artesian wells are drilled at places where the permeable rock layer-like substances receive water along the outcrops at a higher level than the level of the ground surface. The hydrostatic pressure forces the water to the surface and the steady upflow is maintained by the continuing penetration of water into the aquifer at the intake area (USDIBR 1977).

In an artesian well a positive pressure is created by the sandwiching of a porous stone in between two layers (top and bottom) of impermeable rocks like shale or clay (Fig. 1.2). When a permeable stratum is confined between two impervious strata at the top and bottom, artesian conditions exist. The outcrop of the permeable stratum should be at a height enough to produce sufficient hydrostatic pressure on the water at lower points. When a bore is drilled at a proper position through the upper impermeable strata the water rises under pressure through the bore and when the pressure is more water even overflows at the surface.

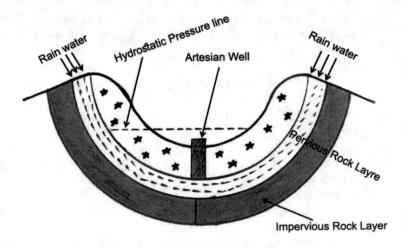

Fig. 1.2 Artesian well

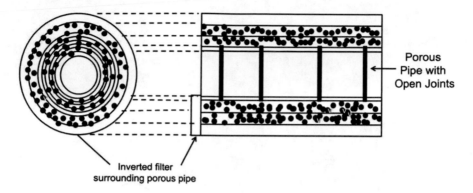

Fig. 1.3 Infiltration gallery

Infiltration Gallery

A structure including perforated conduits in gravel to expedite transfer of water to or from soil is known as infiltration gallery. Infiltration galleries which vary in size, from few meters to many kilometers are usually located close to ponds or streams and may be under the direct influence of surface water (Fig. 1.3). Often and in combination with other water supply systems, an infiltration gallery acts as a means of increasing the quality of water in poor water yielding areas. For this, one or more galleries are built to drain into a central point, such as spring box or hand dug well, known as a collector well.

Infiltration Wells

Also known as interception wells, infiltration wells are the structures that allow surface runoff to drain through underground piping. Unlike drainage wells and inlet wells, infiltration wells do not have a direct water inlet at ground level, rather they increase the infiltration capacity of the ground through the installation of porous materials and, in most cases, a coiled drain between the soil surface and the underground piping (Fig. 1.4). The infiltration wells can be joined to vertical collecting wells or jack wells sunk on the banks of the river, by means of horizontal underground porous pipeline.

While carrying out the sinking operations, it is very essential to see that infiltration well does not tilt as it may cause breakage of pipelines and is also essential to see that after sinking of well there is no appreciable settlement.

Global Distribution of Water

Based on different calculation, the estimates regarding the global distribution of water resources have produced varied scenarios. It is estimated that the total volume of water on earth is approximately 1.4 billion km^3 (Dingman 2002). Out of

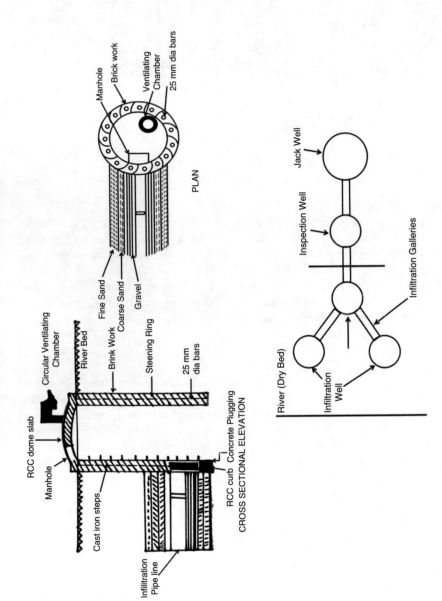

Fig. 1.4 Infiltration wells

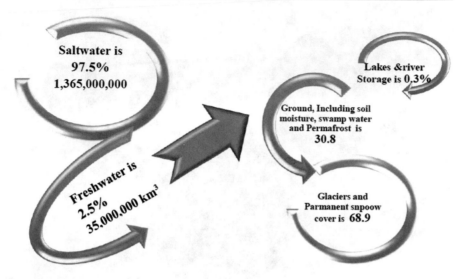

Fig 1.5 Global water distribution

this amount only $8.336 \times 10^6 \, km^3$ (approximately 0.68% of the total water of earth) is stored in the form of groundwater, (Das 2006) and approximately $4.168 \times 10^6 \, km^3$ in the upper 0.8 km of the earth's crust with the rest $4.168 \times 10^6 \, km^3$ lying in the deep strata (beyond 0.8 km from the ground level). Approximately 97.5% of the global water is salty and 2.5% is fresh (Fig. 1.5) with 68.7% of the fresh water present as glaciers, 30.1% as groundwater, 0.8% as permafrost, and 0.4% as surface and atmospheric water (CGWB 2006). Out of the 0.4% surface and atmospheric water, 67.4% is present in freshwater lakes, 12.2% is present as soil moisture, 9.5% is present as atmospheric water, 8.5% is present in wetlands, 1.6% in rivers, and rest (0.8%) as biological water (UNEP 2002).

Groundwater Availability in India

As per Ministry of Water Resources, Govt. of India, the total annual replenishable groundwater of the country is estimated as 433 billion cubic meter (BCM) with around 34 BCM as the natural outflow. The net annual groundwater resource of the country is around 399 BCM and the annual withdrawal of groundwater is estimated as 231 BCM (Foster and Mandavkar 2008), out of which 213 BCM is used for irrigation purposes and rest is withdrawn for domestic purposes (Fig. 1.6).

Groundwater Balance

Water enters into an aquifer primarily through the process of recharge from rainfall. It may also occur by the processes of recharge from canal seepage (R_r), return flow from irrigation fields (R_f), leakage from overlaying and underlying aquifers, i.e., leaky aquifers (Q_{li}), artificial recharge (Q_r), seepage from streams and lakes (Q_{si}), inflow from the neighboring basins (Q_i). However, it may come out from an aquifer by the process of withdrawal (Q_p), evapotranspiration (E_t), outflow to the

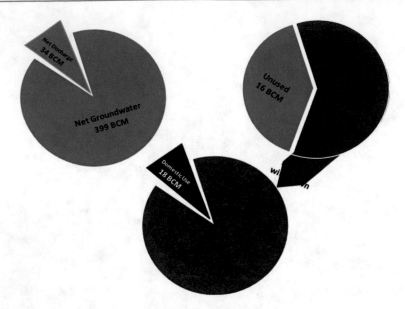

Fig. 1.6 Groundwater resources of India

neighboring basins (Q_o), seepage to the streams and lakes (Q_{so}), leakage to overlaying and underlying aquifers (Q_{lo}), and discharge through spring (Q_s). So, considering these inflow and outflow options the groundwater balance equation can be written as:

$$R_r + R_f + Q_{li} + Q_r + Q_{si} + Q_i = Q_p + E_t + Q_o + Q_{so} + Q_{lo} + Q_s + \Delta S$$

where ΔS is the change in storage.

However, due to some practical impossibilities it is impossible to calculate all the components of this equation, so the components are lumped together to get their net response.

Conflicts Over Water

Water conflict is a term describing a conflict between countries, states, or groups over access to water resources (Wolf et al. 2003). The United Nations recognizes that water disputes result from the opposing interests of water users. Water has historically been a source of tension and a wide range of water conflicts appear throughout history, though rarely are traditional wars waged over water alone. The water conflicts arise for several reasons, including territorial disputes, a fight for resources, and strategic advantage (Gleick 1993). A comprehensive online database of water-related conflicts (the Water Conflict Chronology) has been developed by the Pacific Institute and the conflicts have been found to occur over both freshwater and saltwater, between and within nations. However, conflicts mostly occur over freshwater; because freshwater resources are critically important, yet scarce. They are the center of these disputes arising out of the need for potable water, irrigation, and energy generation (Barnaby 2009). As freshwater is a vital, yet unevenly distributed natural resource, its availability often impacts the living and economic conditions of a country or region. The lack of cost-effective water supply options in areas like the Middle East, among other elements of water crises can put severe pressures on all water users, whether corporate, government, or individuals, leading to tension, and possibly aggression. Recent humanitarian catastrophes, such as the Rwandan Genocide or the war in Sudanese Darfur, have been linked back to water conflicts. A recent report "Water Cooperation for a Secure World" published by Strategic Foresight Group concludes that active water cooperation between countries reduces the risk of war. This conclusion is reached after examining trans-boundary water relations in over 200 shared river basins in 148 countries (Gleick and Heberger 2012).

Causes

According to the 1992 International Conference on Water and the Environment, water is a vital element for human life, and human activities are closely connected to availability and quality of water. Unfortunately, water is a limited resource and in the future its access "might get worse with climate change, although scientists projections of future rainfall are notoriously cloudy," writes Roger Harrabin. Moreover, "it is now commonly said that future wars in the Middle East are more likely to be fought over water than over oil," said Lester R. Brown at a previous Stockholm Water Conference.

Water conflicts occur because the demand for water resources and potable water exceeds supply, or because control over access and allocation of water are disputed. Elements of a water crisis may put pressures on affected parties to obtain more of a shared water resource, causing diplomatic tension or outright conflict. Eleven percent of the global population, or 783 million people, are still without access to improved sources of drinking water which catalyzes the water disputes (Annan 2002). Besides life, water is necessary for proper sanitation, commercial services, and the production of commercial goods. Thus numerous types of parties can become implicated in a water dispute, e.g., corporate entities may pollute water resources shared by a community, or governments may argue over who gets access to a river used as an international or interstate boundary. The broad spectrum of water disputes makes them difficult to address. Local and international laws, commercial interests, environmental concerns, and human rights questions make water disputes complicated to solve—combined with the sheer number of potential parties, a single dispute can leave a large list of demands to be met by courts and lawmakers.

Examples of Water Conflicts

International
1. The Indus between India and Pakistan
2. The Colorado River between Mexico and the United States
3. The Shatt-al-Arab between Iran and Iraq
4. The Brahmaputra between Bangladesh and India

National
1. Sharing of Cauvery water between Tamil Nadu and Karnataka
2. Sharing of Krishna water between Andhra Pradesh and Karnataka
3. Sharing of Siruvani water between Kerala and Tamil Nadu

Management of Water Conflicts

1. To enforce/implement laws that check the water pollution.
2. To overcome the problem of river water sharing in a country, interlinking of rivers should be done.
3. To ensure equitable sharing of basin water, the river basin authority and natural water authority should be given full powers.
4. Nationalization of rivers should be done.

Conservation of Water

Putting the water resources for the best beneficial use with all the technologies available at our command is known as water conservation. It refers to the reduction in the recycling of waste water and usage of water for different purposes such as manufacturing, cleaning, irrigation, and agriculture or else it can be defined as a behavioral change or it can be a technological device or process or an improved design implemented to reduce water loss or use (Arceivala and Asolekar 2007; CWC, Report 2005).

Need

1. It is needed to restore the fast deteriorating aquatic ecosystems.
2. To meet any inevitable future emergency of water shortage for drinking and domestic purposes.
3. To match the demand and supply of clean waters for the present and future generations.

Approaches of Water Conservation

Social

- Initiation of water conservation programs at local level, by the municipal bodies or the regional governments
- Start up of public outreach campaigns and strategies to restrict the outdoor water use for lawn watering and car washing
- Universal water metering to increase the efficiency of the water system
- Proper monitoring of water usage by public, domestic, and manufacturing services to ensure less wastage of water

Domestic

- Low-flow shower heads and toilets flush tanks
- Use of dual flush toilets as they use upto 67% less water than the conventional toilets
- Use of harvested rainwater for toilet flushing
- Usage of faucet aerators, which break water into fine droplets to maintain wetting effectiveness
- Recycling and reuse of wastewater
- Harvesting of rainwater
- Use of high efficiency cloth washers
- Use of weather-based irrigation controllers
- Use of low flow taps in wash basins and use of automated nozzles

Commercial

- Use of infrared or foot-operated taps, which can save water by providing short bursts of water for rinsing in a kitchen or a bathroom
- Use of pressurized water brooms, instead of a hose to clean sidewalks
- Use of water-saving steam sterilization systems for hospitals
- Rainwater harvesting at commercial scale

Agricultural

- Optimization of efficient irrigation systems to minimize losses due to runoff, subsurface drainage or evaporation and to maximize production
- Increasing the efficiency of existing water supply systems
- Replacement of flood irrigation by sprinkler/drip-type irrigation systems
- Use of water-efficient varieties of crops
- Preparation of land for efficient water usage

- Use of rainfall and soil moisture sensors to optimize irrigation schedules
- Measurement and more effective management of the existing irrigation systems

Watershed: Concept, Approach, and Framework

Watershed, a natural hydrologic entity encompasses a specific area/stretch of land where from runoff or rainfall flows into a specifically defined drain be it a nullah/ channel, river or small stream. A watershed is an area of land that feeds all the water running under it and draining it off into a water body. It combines with other watersheds to form a network of rivers and streams that progressively drain into larger water areas. Watersheds are considered as appropriate units for both surveys and assessment of soil and land resources as well as for planning and implementation of various developmental programs.

The process of delineation of watersheds has scientific and rational approaches. However, for presenting them as units for planning and development, formulation of an appropriate framework is a prerequisite. Such framework shall not only follow a hierarchical system of delineating bigger hydrologic units into watersheds but should also include system codification so that each watershed or units of intervention can be identified as a "unique" individual entity with clearly identifiable linkage with other higher hierarchical units, like basin, catchments, and sub-catchments (Kakade and Hegde 1998). However, framework of watersheds requires a different approach indicating macro and micro level delineation so that programmatic intervention units are of manageable sizes (Russell et al. 2014).

Watershed Characteristics

The main characteristics of a watershed include:

- Size
- Shape
- Physiography
- Climate
- Drainage
- Land use
- Vegetation
- Geology and soils
- Hydrology
- Hydrogeology
- Socioeconomics

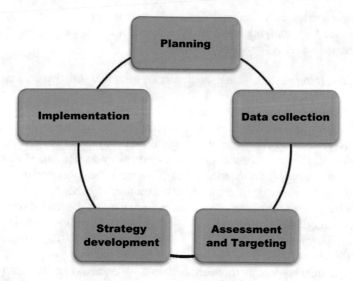

Fig. 1.7 Steps of watershed approach

Watershed Approach

The watershed approach is a cycle of tasks including setting of standards for surface water quality, measuring the water conditions, assessing the data, identifying impairments, establishing priorities, verifying the pollution sources, developing water quality restoration plans, implementing control strategies for various pollution sources including permits, rules, and management practices for nonpoint sources (Russell et al. 2014). Various steps involved in the watershed approach are (Fig. 1.7):

- *Planning*: It involves the identification of different resource personals and stakeholders for watershed planning unit.
- *Data collection*: It involves the collection of routine water quality and quantity data at specific locations.
- *Assessment and targeting*: It involves comparison of the current water quality to the state and federal standards.
- *Strategy development*: It involves the development of strategies and goals to achieve or maintain water quality standards and to meet future demands.
- *Implementation*: It involves implementation of strategies and goals through best management practices (BMPs), permits, and education.

Framework of Watershed

Nine essential elements including geographic management units, stakeholder involvement, basin management cycle, strategic monitoring, basin assessment, priority ranking and resource targeting system, capability for developing management strategies, management plan documentation and implementation are recommended for a unifying watershed management framework.

Specific steps to this framework include:

1. Planning Determine the watershed planning unit and identify the stakeholders and resource personals.

Determination of the Watershed Planning Unit

Varying from mainline portions of large river basins to small areas discharging excess flows into ponds, the size of watershed units influences the role of different stakeholders in all steps of watershed management cycle (Paranjape et al. 1998). Although the geographic boundaries usually determine the watershed units they may also be defined according to the authority exercised by the governments over a particular land area as for example a local agency may be the lead stakeholder in a sub-watershed of 1–10 square miles (Russell et al. 2014) while as a federal or state agency may play a larger/lead role in a large river basin of 100–10,000 square miles.

Identification of Resource Personal and Stakeholders

Although stakeholders, both direct (within a watershed) and indirect (outside a watershed) may vary within a defined watershed management, a representative mix of stakeholders is critically important to the successful management of such units.

Resource personals are individuals or groups of individuals who bring specific technical expertise to the decision makers and stakeholders. They may include engineers, scientists, conservationists, attorneys, or policy experts that can advise stakeholders on questions and specific issues. However, in certain cases the resource personals may be contracted from the private sector while in certain other cases they may be a part of the stakeholder group including landowners (permanent and absentee), homeowners, local businesses, agricultural producers, industry, city and county officials, water and wastewater utilities, environmental activists, conservationists, civic groups, and mass media activities.

2. Data Collection As the degradation of water quality or limited supply of water is the main reason for the development of a management plan for a watershed, the data collection phase involves a routine work of the collection of water quality and quantity data at specific sites. The point and nonpoint sources which in totality are responsible for the total pollution load are the sources that contribute to water quality.

Point source pollution which comes from confined sources like wastewater discharge points, storm water collection systems and industries etc. are typically monitored and regulated for their quality standards by a federal or state agency. The nonpoint source pollution comes from different sources like:

- Herbicides, insecticides, and fertilizers from residential areas and agricultural lands
- Toxic chemicals, greases, and oils from energy production systems and urban runoffs

- Sediments from improperly managed croplands, construction sites, and eroding stream banks and forest lands
- Salts from irrigation practices
- Acid drainage from abandoned mines, bacteria, and nutrients from livestock, wastes, and faulty septic systems

Although costly at times, implementing tougher standards at point sources is typically easier because the pollution is delivered to one area for treatment; however, for nonpoint sources it is generally more difficult to control because it is not centrally collected and can be a result of numerous factors which again are not always specific to land areas adjacent to streams and could be a result of secondary impacts.

3. Assessment and Targeting The next step in the watershed management approach involves the comparison of current water quality to the federal or state water quality standards.

Standards Designed to Assessment and Targeting

Numerical and narrative goals for water quality can establish reasonable methods to attain and implement the state's goals for water quality. All standards are protective; that is, they signal a situation where there is some possibility that water quality may be inadequate to meet its designated uses. There are instances, for example, in which a water body fails to meet the standards for aquatic life, yet no fish kills are observed. However, a decline in the number or variety of aquatic species and an increased probability of fish kills may be observed. Aquatic life use, public water supply, contact recreation and fish consumption are the four general categories defined by TCEQ (Texas Commission on Environmental Quality), for surface water quality standards.

Aquatic Life Use

The standards which establish optimal conditions for the support of aquatic life and define indicators used to measure whether these conditions are met or not are designed for the protection of plant and animal life in and around the water bodies. Low levels of dissolved oxygen, or high levels of toxic substances such as pesticides or metals are some conditions that may violate these standards.

Contact Recreation

These standards measure the level of certain bacteria in water to estimate the relative risk of different water sports like swimming, involving direct contact with such waters.

Public Water Supply

These are the standards which are associated with the checking of the safe and unsafe nature of the public water supply systems. Indicators used to measure the safe and unsafe nature of surface water bodies include the presence or absence of substances like pesticides or metals, and some salts like sulfates and chlorides.

Fish Consumption

These standards are designed to prevent the public consumption of fish or shellfish that may be contaminated by different pollutants. They identify levels at which there is a significant risk of accumulation of a toxic substance in the tissue of the fishes and other aquatic species.

As a result, the state also conducts tests on fish and shellfish tissues to determine if there is a risk involved in their consumption or not. Furthermore, to ensure the safety of oysters or other shellfish from the accumulation of bacteria for their commercial exploitation, these standards also specify bacterial loads in marine waters.

4. Strategy Development It involves the development of strategies and goals to maintain, or achieve water quality standards to meet the future demands. This is the key step at which both direct and indirect stakeholders become key players in the identification of strategies and designation of actual watershed management plans, which at times involves experts, policy makers, and interest groups including private and public engineers and scientists so that the decisions can be made with full access to practical and policy implications. Public policy participants may include local, regional, state, and federal personnels including public or private engineers and scientists who may be responsible or may have interest in management of watershed plans.

Outlining specific goals enables the stakeholders to achieve the required results in an easy manner. Water quality models are tools that allow users (engineers, planners, managers, etc.) to mathematically simulate natural processes in a watershed using a personal computer. Models generally require information on topography, land use, climate, and soils so as to assist stakeholders in evaluating the impacts of various management strategies and land use changes on the watershed. Although models can be a very useful tool, they have limitations due to scale (size of the watershed) and available data (water quality parameters, stream flow, etc.). In addition, modeling efforts need to be combined with social acceptability to achieve successful results.

5. Implementation It involves the implementation of strategies and goals through best management practices (BMPs), education and permits, which are the new tools to put the watershed management plans into practice. However, the stakeholders can customize their tools for putting the watershed management plant to best use in their own best way.

Permits

Regulatory permits specifying discharge level of pollutants issued by the government are often used to control the point sources so that they do not exceed the permissible limits. A management plan that uses these permits as the main tools of management may be effective only if point sources are the main contributors to water quality.

Best Management Practices

For the management of nonpoint pollution sources, BMPs is the preferred approach. Although voluntary in nature BMPs are used by some regulatory agencies in the watershed management plans, e.g., NPDES (National Pollution Discharge Elimination System) combines construction permits with the BMPs to control the runoff and soil erosion.

Educational Programming

Education, a key to successful watershed management plans alerts the stakeholders about the watershed problems and involves them in the decision-making process. It also draws the attention of both agency stakeholders and employees to the need of proper strategic balance between BMPs and permits which leads to the management plans that address pollution from both point and nonpoint sources (Russell et al. 2014).

Degradation of Water Bodies of Kashmir and Their Management

Deterioration of environmental quality through degradation of ecosystems, depletion of resources, loss of biodiversity, and loss of habitats is known as environmental degradation or it may be defined as the deleterious and undesirable changes brought about in the environment with depletion of fresh water resources as one such deleterious change (Rani 2016). Estimates say that over the entire globe, one in three people are already facing water shortage, almost one-fifth of the world population lives in areas of physical water scarcity and one quarter of the population living in the developing world lack infrastructure to use water from aquifers and rivers etc. This water shortage/crisis is expected to deepen further in the wake of some foreseen issues like population explosion, urbanization, industrialization, increased incidences of climate change, and higher standards of living.

The Valley of Kashmir has over a thousand small and large water bodies, which are the bedrock of both its ecology and economy. But unfortunately over the last century, massive urbanization around these water bodies has led to pollution, siltation, and overexploitation of many streams and lakes (Mushtaq et al. 2018) and due to such activities many have shrunk to a fraction of their original size while some have completely disappeared and some are on the verge of extinction.

According to the State Water Mission, the water bodies in Kashmir are the worst victims of increased urbanization through human interference. Large-scale deforestation in their catchment responsible for soil erosion results in their siltation and thereby their conversion into landmasses. While some water bodies have completely disappeared through the natural acts of low precipitation and glacial action.

During the past century almost above 50% of water bodies in Srinagar and its suburbs have been lost as for example the increased deforestation in the Jhelum basin has led to an excessive siltation of most of its water bodies especially lakes, a crisis that is further deepened by the subsequent human greed brought about by change in land-use pattern around these assets of high ecological value. More than 9119 hectares of wetland and open water surface have disappeared between 1911 and 2014, while only 6873 hectares have been preserved.

The marshy and watery area of Dal Lake, a major tourist attraction in Srinagar, has shrunk from 2547 hectares in 1971 to 1620 hectares in 2008 meaning it has lost almost half of its water surface area in a gap of 40 years, although it still looks like a water body. Adjoining lakes like Gilsar, Aanchar, and Khushalsar have all but disappeared because the drainage system of the Dal which used to feed these small water bodies has been converted into landmass through heavy siltation. Dal Lake has not only been a center of civilization and a beautiful national heritage but has played a major role in the economy of the state through its attraction of tourists as well as its utilization as a source of food and water. During the past few years grave concern is being voiced by many researchers and people from different walks of life over the deteriorating conditions of this lake and there is no deny in the fact that the lake has fallen prey to the increased human greed. A large area of the lake has been cultivated and converted into floating gardens. Over the years the lake has become shallow due to siltation and accumulation of plant debris as the lake is under the influence of different anthropogenic pressures. The increasing human activities in the four basins of Dal Lake are mainly responsible for the deterioration of sediment and water quality and decline in benthic fauna in the lake. The cumulative impacts of their direct as well as indirect drivers have created a number of pressures on the ecology of the lake (Mushtaq et al. 2018).

Wular Lake, one of the largest fresh water lakes of Asia plays an important role in the hydrographic system of Kashmir valley and acts as a huge absorption basin for the annual flood waters. The lake with its extensive surrounding marshes is the natural habitat of a variety of wild life. Accounting for about 60% of the state's total fish production, this lake is a source of sustenance for a huge chunk of population living along its fringes and based on its hydrological, socioeconomic, and high biological value, it has been declared as a wetland of national importance in 1986 by the Ministry of Environment and Forests, under its wetland program leading to its international importance by its subsequent declaration as Ramsar Site in 1990 (Anon 1990). But with regards to its adjacent/associated marshes there is a remarkable alteration in its area as it has got reduced by more than 41 km^2 during the past 100 years. As per the Survey of India map of 1911 the open water area of Wular Lake was 91.29 km^2 which got reduced to 79.82 km^2 in 1965 (Mir 2006) and presently as per LISS+PAN merged satellite image of 2007 the open water area of the lake is 75.23 km^2. Similarly, the wetland area surrounding the lake body and in the adjacent area was 66.45 km^2 and 58.67 km^2, respectively (Survey of India Map 2011) out of which we have lost 54.97 km^2 in and around the Lake and 41 km^2 in its surrounding marshes to Horticulture/Agriculture and plantation etc. during the past 100 years.

Management

Keeping in view the following objectives the government shall prepare the State Water Policy and Plan for the management, development, utilization, planning, and monitoring of the state water resources:

1. The availability and demand of water for various purposes like agriculture, domestic, industry and power etc.
2. The obligations of the state under any treaty, agreement, order, or judgment of any court or tribunal, or statutory obligations under any law for the time being in force in the state
3. The management and conservation of available water resources in most economical and sustainable manner in keeping with the ecological and environmental concerns
4. Scientific management of floods and droughts
5. Involvement of different socioeconomic parameters like environmental sustainability, rehabilitation, and resettlement of project affected people and livestock, in the planning and management of multipurpose river valley and irrigation projects
6. Improvement and development of surface and groundwater by preventing its overexploitation and ensuring its sustainability
7. Proficient water utilization for different purposes to promote the conservation and augmentation of traditional water resources
8. Usage of scientifically innovative techniques for improvement of water quality and reduction of surface and groundwater pollution
9. Training and capacity building of people involved in the development and management of water resource
10. Utilization of efficient irrigational facilities in agriculture sector and replacement of water intensive crops by those requiring less water for an increased productivity
11. Development of a database of the water availability, usage, future demands, and other hydraulic data with special focus on the use of modern techniques like remote sensing and computer programs to establish a network of data banks for the effective exchange of information among different agencies
12. Development of an effective institutional mechanism for coordinating the management of water resources with a multi-sectorial and multidisciplinary approach
13. Water retention maximization and water loss minimization measures by watershed management through extensive catchment area treatment, soil conservation, wetlands, and preservation of forests increasing the construction of check dams, forest cover, and other groundwater recharge measures
14. Utilization of some nonconventional practices like artificial recharge of groundwater and traditional water conservation practices like rainwater harvesting by roof top rainwater harvesting and encouragement of recycling and reuse of water

Eutrophication

Eutrophication is the enrichment of water bodies with plant nutrients and precursors, typically nitrogen, phosphorus, and organic matter. There exists a natural and slow eutrophication, which over geological times turns a lake into a marsh and then dries it entirely. The term eutrophication which comes from the Greek word "eutrophos" meaning well-fed, is a process of nutrient enrichment of any water body. The definitions of the concept exist in the literature, with some fundamentally different from the others in respect to whether it is only the process of nutrient enrichment or it should include the problems associated with such enrichment (Garg et al. 2002). It results in an excessive growth of aquatic vegetation including phytoplanktons whose undesirable overgrowth and subsequent death forms a greenish slime layer over the surface of the water body thus blocking the penetration of light, restricting the reoxygenation of water through air currents and producing the foul smelling gases and making the water body more turbid. It is one of the serious water pollution problems that due to dissolved oxygen loss directly affects the flora and fauna of these aquatic systems and hence an early and relatively higher mortality rate of fishes. It further makes the fishing operations and navigation in such water bodies difficult due to heavy and enmeshed growth of aquatic plants (Wang et al. 1999). The hydroelectric generation from such water storages is also adversely affected as nutrient-rich water acts chemically upon the turbines. Further, the decomposing debris of the algal blooms also spoils the desired water characteristics and results in an increased growth of disease-causing bacteria. Uncontrolled eutrophication further leads to a rapid upwelling of a water body, reduced/limited water recharging and storage capacity of smaller freshwater bodies by silting (Likens 1972). Ponds and small lakes lose their aquatic identity by becoming permanently terrestrial in nature. It lowers the values of surface water for different purposes while leading to significant changes in quality of water. Besides being dependent upon the supply of nutrient inputs, eutrophication is greatly influenced by a number of environmental factors. Eutrophication induces significant changes in the biodiversity which directly affects the trophic structure of the aquatic ecosystems. Some effective control measures like biological control, mechanical control, legislative measures, and the awareness programs pertaining to the present threat to water resources on the blue planet need to be carried out effectively.

It is a process that occurs naturally in all the aquatic systems and takes thousands of years. However, at a higher rate of nutrient input due to anthropogenic activities it occurs significantly in a very short period of time hence known as artificial/ accelerated eutrophication. Eutrophication causes an increase in plant and animal biomass, frequency of algal blooms, growth of rooted plants, and decrease in the species diversity resulting in an increased turbidity and anoxic conditions. Because of the high density of aquatic organisms in a eutrophic system, there is often a lot of competition for resources, which along with high chemical or physical stress makes the struggle for survival in eutrophic systems far higher (Liu et al. 2004).

The cultural eutrophication consists of a continuous increase in the contribution of nutrients, mainly nitrogen and phosphorus (organic load) until it exceeds the self-purifying capacity of the water body (Antoniades et al. 2011) thus triggering structural changes in the waters which mainly depends on three factors:

- *Use of fertilizers*: Excessive use of fertilizers during agricultural activities results in an accumulation of nutrients which upon reaching the higher concentration levels go beyond the assimilation capacity of soil and are hence carried by rain into rivers and groundwater that flow into lakes or seas.
- *Discharge of wastewater into water bodies*: In various parts of the world (particularly the developing countries) the wastewater discharged directly into water bodies such as rivers, seas, and lakes results in the release of a higher quantity of nutrients which stimulates the disproportionate growth of algae. In industrialized countries, however, the wastewater can illegally be discharged directly into water bodies.
- *Reduction of self-purification capacity*: Over the years, lakes accumulate large quantities of solid materials (sediments) and these sediments absorb large amounts of nutrients and pollutants. Consequently, the accumulation of sediments starts the basin filling and, increases the interactions between water and sediment, thus facilitating the suspension of nutrients present at the bottom of the basin which further lead to the deterioration of water quality.

Global Scenario

Eutrophication has become a major water pollution problem both in the developed as well as developing world as the water bodies of the high pollution nations like India, China, Pakistan, Bangladesh, and Indonesia, the great lakes of Canada and the United States, and industrialized countries of Europe are under the direct threat of eutrophication. Population pressure coupled with unprecedented industrial growth, change in lifestyle and economic development during the recent past has added to this problem (Ansari et al. 2011). During the past 25–30 years there has been a constant rise in the scientific interests in eutrophication which has risen further in the recent past and it is predicted that acidification, eutrophication, and contamination by toxic substances are likely to increase further as a threat to the freshwater ecosystems. Eutrophication was recognized as a pollution problem in North American and European reservoirs and lakes in the mid-twentieth century and since then, it has become more widespread as a survey showed that 28% in Africa, 41% in South America, 48% in North America, 53% in Europe, and 54% of lakes in Asia, are found in an eutrophic state (An 2003). Human-induced eutrophication degrades freshwater systems worldwide by reducing water quality and by altering the ecosystem structure and function. Comparison of the total phosphorus and nitrogen concentration for the USEPA nutrient ecoregions with the estimated reference conditions show an unprecedented increase in all these values for rivers and lakes over the reference median values and it is observed that currently almost over 90% of rivers exceed the reference values.

Due to eutrophication, the shallow brackish lake Hickling Broad (Norfolk, USA) showed changes from a clear macrophytic stage to a turbid phytoplankton dominated stage from the decades of 1960s to mid-1970s which again showed some recovery in terms of submerged plants in 1980s. Likewise the discharge from various point sources like sewage water treatment plants containing huge quantities of nitrogen and phosphorus into Lake Tohopekaliaga, a part of Kissimmee river system in central Florida resulted in a deterioration of its habitat qualities, water qualities and respective biocommunities with the overall phosphorus loading of 11 times higher than the natural conditions in 1979 (Wetzel 2001).

Due to the discharge of huge quantities of domestic and industrial sewage from the entire Lerma-Chapala basin into Lake Chapala (Mexico), the third largest American lake, significant changes lead to a eutrophic condition. The shallow urban city park Lake (Baton Rouge, Louisiana) has also been reported as hyper-eutrophic which suffered from frequent algal blooms and fish kills. Lake Apopka, a large, shallow lake in Florida (USA) was made hyper-eutrophic due to excessive phosphorus loading from the flood plain farms and was reported to have higher levels of nutrients, phytoplanktons and suspended matter. Similarly, the Lake Victoria (Kenfa) has undergone a drastic change during the latter half of the last century with a population decline of endemic fishes as the most noticeable. A dimictic Swiss Lake (Greifensee), which once was oligotrophic in nature turned out to be hypertrophic in early 1970s due to increased phosphorus influx from the heavily populated catchment area. Deterioration in habitat conditions and reduction in phytolittoral zones was observed in a mesotrophic Polish Lake (Mikolayskie) over a period of 30 years due to the increased human interference.

Indian Scenario

India is no longer an exception, when it comes to eutrophication of water bodies, as every small and large water body here in India as well is seen under a direct influence of eutrophication as for example in Lake Mirik, located in the Indian Himalayas the higher influx of nutrients in some of its pockets due to the anthropogenic nutrient inputs has shown a spoiled water quality. Similarly, the discharge of sewage and industrial discharge in river Ganga characterized in terms of its high COD, BOD, hardness, chloride, alkalinity, specific conductivity, phosphates, nitrates, low dissolved oxygen, free CO_2, and pH has shown a quiet large deterioration (Bandela et al. 1999). Lake Dalavayi, Mysore due to additional inputs of soap and detergents has shown a higher metal toxicity and concentration of heavy metals exceeding WHO standards. Bellandur Lake, one of the major lakes of Bangalore in India showed a characteristic alteration from a natural oligotrophic to an artificial reservoir of domestic sewage and industrial effluents by the addition of effluents from city drainage. A study conducted in three lakes (upper lake, lower lake, and Mansarovar reservoirs) in Bhopal depicted a higher level of eutrophication due to direct sewage dumping and idol emersion which significantly boosts their nutrient dynamics (Vyas et al. 2006). Studies on eight shallow lakes, including Gambhir Bandh, Salon, Devrishi Deval, Ratoi, Majuwajagat, Chandu, Bhagnaija, and Bandsideh, located in Uttar Pradesh (India) showed that all these (small lakes) were diatom rich, eutrophic, and mildly to heavily polluted. In Kashmir, world famous Dal Lake has been heavily polluted by different organic pollutants. Direct sewage drainage from

Srinagar city, runoff from floating gardens and nearby agricultural catchment into the Dal Lake, turned out to be a serious threat to the lake in terms of eutrophication (Mushtaq et al. 2013). Similarly, Aanchar Lake in Srinagar Kashmir has also been a victim of nutrient enrichment which resulted in the hypereutrophic conditions of the lake and various recent works have reported that Aanchar Lake has turned into a dystrophic system. In Kashmir valley rural lakes such as Wular Lake and Mansbal Lake have also been reported to be eutrophic by various researches and reported that these two lakes changed their trophic status toward eutrophic primarily due to excessive agricultural runoff and direct dumping of organic sewage not only into the lake systems but their inlets throughout their catchments (Pandit and Yousuf 2002).

Causes

Eutrophication, the undesired increase in the concentration of nutrients in an aquatic ecosystem occurs through its entry from the point and nonpoint sources of pollution. The point source pollution enters a water body from a single identifiable point which is a fixed facility or a stationary location e.g., an industrial plant, a fish farm, or a sewage treatment plant and the nonpoint source pollution enters a water body with no specific point of entry or discharge e.g., fertilizers from the agricultural lands and losses from the atmospheric depositions.

The process of nutrient enrichment can be either of a natural origin or can often be dramatically increased by human activities (cultural or anthropogenic eutrophication). Natural eutrophication which is a process of addition, flow, and accumulation of nutrients to water bodies resulting in changes to the species composition and primary production has been occurring for millennia. However, cultural eutrophication is a process that is speeded up by the different types of human activities with sewage from cities and industrial wastewater, erosion and leaching from fertilized agricultural areas, and atmospheric deposition of nitrogen (from animal breeding and combustion gases) as the three main sources of anthropogenic nutrient influx (Chislock et al. 2013).

Nitrogen and phosphorus entering the aquatic ecosystems via surface water, groundwater, and air are the most common nutrients with agricultural land as the main source of nitrogen and households and industries as the main source of phosphorus. The rise in intensive fertilizer use has serious implications for the coastal habitats and the fraction of fertilizer lost from fields increases with the intensity of application. Increased global production of nitrogenous fertilizers has largely been linked to concerns over the relationship between eutrophication and water quality.

Natural eutrophication depends only on the natural features and local geology of the catchment. However, cultural eutrophication is associated with human activities which accelerate the process of eutrophication beyond the rate associated with the natural process.

Control Measures of Eutrophication

Mixing and Oxygenation

The aim of artificial mixing procedures is either to oxgenate the deoxygenated hypolimnion or the entire water body and/or inhibit the phytoplankton growth. Destratification is accomplished by injection of compressed air from a diffuser into the reservoir bottom with the following three goals:

1. Destratification, to prevent algae from remaining in the illuminated layer and causing a decrease in phytoplankton biomass formation
2. Circulation, to cause a shift from blue green to less noxious green algae
3. Aeration, to oxidize the hypolimnion and consequently seal the bottom to prevent release of manganese, iron, and phosphorus

Destratification has the following advantages:

1. Increase in hypolimnetic oxygen
2. No or low phosphorus release from the sediments
3. No or low amounts of manganese and iron
4. Decrease in the amount of algae

Layer aeration is a relatively new approach that is based on detailed knowledge of stratification conditions in a given water body. Here the oxygen and heat in a stratified water body are redistributed into discrete layers and the manipulation of the thermal structure creates desirable physical and chemical conditions to avoid the negative effects of destratification.

Methods of Treating Sediments

Over longer periods of time the sediments accumulate phosphorus in the upper few millimeters of the sediment than the phosphorus content in the entire water column and the dissolved fraction of this huge phosphorus store is constantly exchanged with the overlying water column. Because of the large phosphorus storage in the sediments, eutrophic conditions continue for several years after phosphorus supply to the reservoir is considerably reduced. Various procedures are used to decrease the release of phosphorus from the sediments.

Sediment removal consists of removing the upper layers of sediment that contain high phosphorus levels. With long lasting results, sediment removal technique has been used in different eutrophic water bodies for the abatement of eutrophication and to stop this process before any hypereutrophic condition. A good example of sediment removal can be seen in Dal Lake Kashmir where contaminated sediments have been removed by dredging, that not only resulted in the reduction of nutrient levels (particularly nitrates and phosphates) but also in the reduction of shallowness of the lake (Mushtaq et al. 2018).

Biomanipulation

The term "biomanipulation" (Shapiro et al. 1975) is a form of biological engineering in which organisms are selectively removed or encouraged to alleviate the symptoms of eutrophication. Its principle is food chain manipulation, by maintaining low feeding pressure on zooplanktons by fish, so that large species of zooplanktons that are capable of keeping phytoplanktons under control. Development of fish populations that lead to control of zooplanktons and phytoplanktons can be achieved by the following ways:

1. Temporary eradication of stunted fish populations by using rotenone poisoning and predator stocking (rotenone is nontoxic for invertebrates and phytoplankton)
2. Continuous introduction of predatory fish and net harvesting of non-predatory fish; collaboration with local sport fishery and use of commercial fishery methods
3. Reservoir drawdowns during reproduction periods of undesirable fish species by exposing eggs on shore vegetation

The advantages of this method, besides being low cost, are that it is fully natural, with no chemicals or machinery required. Costs are low if combined with organized fisheries efforts but increase when these efforts are not combined.

Rainwater Harvesting

Used as one of the most efficient and effective way of management and conservation of water, it refers to the collection and storage of rainwater at the surface or in the subsurface aquifers for the use of humans, animals, and plants. Rainwater harvesting is a very useful practice of water conservation for countries like India as it can reduce the demand and cost of treated water along with the economization of the maintenance and distribution costs and treatment plant operations.

Need of Rainwater Harvesting
1. It helps us to overcome the less availability of surface water.
2. By replenishing the groundwater it helps us to get rid of the decline in groundwater and enhances the availability of groundwater.
3. For utilization of rainwater in sustainable development.
4. Improvement of the quality of groundwater by dilution through increased infiltration.
5. Improvement of agricultural production and urban ecology by increasing the vegetation cover.

Advantages of Rainwater Harvesting
Advantages of rainwater harvesting are so, that the collection and storage of rainwater runoff on a surface area of merely 2% of India can supply 26 gallons of water per person (Centre for Science and Environment). So, much efforts are made and are

needed to popularize this concept at the grassroot level in order to increase its advantages much beyond the below mentioned advantages:

1. It helps in the reduction of soil erosion, effects of droughts, and the flood hazards.
2. It results in a rise in groundwater table by increasing the productivity of aquifers.
3. It helps in the storage of groundwater in a more environmental friendly manner as the same is not supposed to the risk of evaporation and pollution.

Methods and Techniques of Rainwater Harvesting

The various techniques and methods used for rainwater harvesting vary from urban to rural areas with rooftop rainwater/storm runoff harvesting through recharge pits, recharge trenches, tubewells and recharge wells as the most common method used in urban areas; however, in small areas, rainwater harvesting is done through gully plugging, contour bunding, gabion structures, percolation tanks, check dams/cement plugging/nalla bunding, recharge shafts, dugwell recharges, and groundwater dams/subsurface dykes, etc.

Rainwater Harvesting in Urban Areas

Generally speaking the rainwater available in urban areas from the paved and unpaved areas and rooftops of buildings goes waste; however, it can be used to recharge the aquifers and can be utilized gainfully at the time of need.

In urban areas the rooftop rainwater harvesting technique can be effectively used to fight the problem of water scarcity; however, its cost effectiveness depends upon the availability of rainfall, the rooftop area, and its design so that it does not occupy large space for collection and recharge (Patel and Shah 2008).

Some common techniques used in rooftop rainwater harvesting in urban areas are as follows:

1. Recharge pits:
   Recharge pits, though of any shape and size, are generally constructed with a width of 1–2 m and a depth of 2–3 m filled in graded form by 50–20 cm sized boulders at the bottom followed by 5–10 mm sized gravels and 1.5–2 mm sized coarse sand so that the coarse sand at the tops stops the slit content that comes along the surface runoff (Patel and Shah 2008). However, the same is filled with broken bricks/cobbles for small roof areas. It is suitable in areas where permeable rocks are either exposed or are located at very shallow depths below the ground. Recharge pits are constructed for shallow aquifer recharging and are suitable for buildings having a larger rooftop area of about 100 m^2.
2. Recharge trenches:
   Recharge trenches are the structures which are generally constructed with a depth of 1–1.5 m, width of 0.5–1 m, and a length of 10–20 m, depending upon the availability of water to be recharged. They are again filled in the same fashion in which recharge pits are filled. These trenches are suitable in areas where the permeable strata are available at the shallow depths and the building has a quiet larger rooftop area of 200–300 m^2 (Das 2000). To stop the entry of any solid

waste/debris or fine solids in the recharge pits it is provided with a mesh at the roof and a desalting/collection chamber at the ground. Further, periodic cleaning of the top lying sand layer is needed for its proper maintenance.

3. Tube wells:

This technique is suitable to recharge the deeper aquifers in those areas where the existing tube wells are tapping the deeper aquifers due to the drying of shallow water aquifers. Here in this technique, 10 cm diameter PVC pipes are connected to the rooftop drains to collect the subsequent rain showers after letting go the first shower through the bottom of the pipe. The collected rainwater is taken to a filtration system provided before the water enters the tube well. With its diameter depending on the roof area, its length is generally 1–1.2 m and is provided with a reducer of 6.25 cm on both the sides (Das 2000). The filtration system that is divided into three chambers by PVC screens is fitted in the first by 6–10 mm sized gravels, followed by 12–20 mm sized pebbles and 20–40 mm sized bigger pebbles (Patel and Shah 2008).

4. Recharge wells:

Recharge wells are the structures constructed generally with a depth of 3–5 m below the water level and a diameter of 100–300 m. These structures are suitable for areas where the surface soil is impervious and roof top runoff or surface runoff is suitable as large quantum even in short spells of rainfall are produced. Here the pits/trenches are used to store the water which is subsequently recharged to groundwater through the recharge wells. It is ideally suitable for areas where permeable horizon lies well below 3 m of ground level with the design of the well assembly depending upon the lithology of the area. In case the aquifers are located at greater depths say 20 m, a shallow shafts of 2–5 m diameter and 3–5 m depth is constructed as per the availability of runoff. Inside the shaft, a recharge well of 100–300 mm diameter provided with a filter media to avoid its choking is constructed for recharging the available water to deeper aquifers.

Rainwater Harvesting in Rural Areas

Since plenty of space and recharge water is available in rural area, the surface spreading techniques are very commonly used for rainwater harvesting (Sachin et al. 2018) and of these the most commonly used ones are:

1. Gully Plugs

The structures made by using local clay, stones, and bushes across small streams and gullies running down the hill slopes carrying drainage to tiny catchments during rainy seasons are known as gully plugs. The site of construction of gully plugs depends on the local breaks in the slopes which allow accumulation of adequate amount of water behind the bund. The structures are very helpful in conserving the soil and soil moisture.

2. Contour Bunds

These are the structures which intercept the flowing water well before it attains the errosive velocity and act as an effective method of soil conservation in a watershed for a longer time. These structures are suitable in low rainfall areas where the monsoon runoff can be impounded by constructing the bunds along the slope. Spacing between the bunds in contour bunding depends upon the permeability of soil and the slope of the land with closer spacing between the bunds in the areas of lesser soil permeability.

3. Gabion Structures

Gabion structure is a kind of check dam commonly constructed across small streams (width < 10 m) to conserve stream flows with practically no submergence beyond stream course. In this a small bund made up by putting locally available boulders in a steel wired mesh, enclosed to the stream bank is constructed across a stream and the height of the structures is kept around half a meter. In due course of time the silt content of the stream is deposited in the interstices of the boulders with which the growth of vegetation becomes quite impermeable to help in the retention of surface runoff for enough time to recharge the groundwater.

4. Percolation Tanks

Percolation tanks are artificially created surface reservoirs which submerge the highly permeable land so that percolation of surface runoff and the subsequent recharge of groundwater is made possible. Percolation tanks which are mostly earthen dams with masonry structure only for spillways are preferably constructed on second-order or third-order streams, located on highly weathered and fractured rocks with lateral continuity downstreams. These structures facilitate the recharge of groundwater to benefit the cultivable land from the augmented groundwater.

5. Check Dams/Nalla Bunds/Cement Plugs

These are the dam structures which are created across small streams with a gentle slope at a site with a sufficiently thick weathered formation or permeable bed to facilitate recharge of storm water in a lesser time. Normally with a height of <2 m, such structures are mostly confined to stream courses to allow the excess water to flow over the wall and in order to avoid scouring due to the excess runoff, the downstream side is provided with water cushions. These structure are made either of the clay-filled cement bags arranged as a wall or at places by erecting two asbestos sheets across a nalla and back filling the space by clay to create a low-cost check dam to provide stability to the structure created through the latter way. Clay-filled cement bags are stacked in a slope on the upstream side of the asbestos sheet. A series of such check dams are created for the maximum harnessing of the surface runoff in the stream.

6. Recharge Shafts

One of the most efficient and cost-effective mechanisms to recharge the unconfined aquifers, recharge shafts are manually dug in case the strata are of non-caving nature. With a diameter of <2 m, these structures are very useful for water bodies where shallow clay layer impedes the infiltration of water to the aquifers. The recharge shafts being unlined or lined, help in the recharging of the surplus water in the tanks to the groundwater as in case of lined shafts the recharge water is fed to the filter pack through a smaller conductor pipe and in case of unlined shafts through a filter initially back filled with cobbles/boulders followed by sand and gravel. Sometimes depending on the quantum of water availability, the recharge shafts are created with a diameter of 0.5–3 m and a depth of 10–15 m and the top of the shaft is kept above the tank but level preferably at half of full supply level. This technique helps us to recharge the groundwater by almost 50% of full supply level of the village tanks, while keeping the rest of the water available for domestic use.

7. Dug Well Recharge

Use of the abandoned or existing dug well as recharging structures after their proper cleaning is known as dug well recharge. In this technique the recharge water is guided through a pipe from the desiltation chamber to the bottom of the well or below the water level to avoid scouring of bottom and entrapment of air bubbles in the aquifer. Further, to control any bacteriological contamination in the system periodic chlorination is done (Sachin et al. 2018).

8. Groundwater Dams or Subsurface Dykes

These are subsurface barrier across streams which retard the base flow and store water upstream below ground surface. By doing so, the water level in upstream part of groundwater dam rises, saturating otherwise dry part of aquifer. These structures are erected in sites where there is a shallow impervious layer with wide valley and narrow outlet. PVC sheets or low density polythene films are used to cover the cutout dyke faces to ensure a total imperviousness.

Groundwater Harvesting

Groundwater harvesting is the employment of both traditional/conventional as well as nonconventional/nontraditional methods like Qanat system, special types of wells, underground dams for the extraction of groundwater. With the above mentioned term as a few example of groundwater harvesting, subsurface dams and sand storage dams are other fine examples of groundwater harvesting. These structures obstruct the flow of ephemeral streams in a riverbed and use the water stored in the underlying sediments for aquifer recharging (Oweis et al. 2012).

Qanat System

Qanat is a long conduit or tunnel leading from a dug well at a reliable source of groundwater. The tunnel leads gradually down the slope to the communities in the valley below from a mother well dug in the foothills of a mountain range or at the base of a hill. To allow for the maintenance and construction of the qanats some access shafts are intermittently dug along the horizontal conduit (Oweis et al. 2012). Qanat systems are widely used across the Middle East and Persia for the following reasons:

1. It requires no energy.
2. It relies on the gravitational force alone.
3. Without leakage, pollution and evaporation it causes water to cross long distance through land terrain chambers.
4. It ensures the maintenance of water table as the discharge is fixed by nature, producing only the amount of water that is distributed naturally from a spring or mountain (Adle 2005).
5. It allows access to a plentiful and reliable source of water to those living in otherwise marginal landscapes.

Water Loss: Management to Minimize Loss

Water delivery to the consumers is significantly affected by the water loss from a water distribution system which can either be due to unauthorized consumption and apparently due to meter inaccuracies or real loss due to breaks/leakages on mains and links or at the storage facilities. Awareness about the occurrence of water loss in a system is the first step in identifying leaks and their subsequent repairs. It is only after the identification and documentation of water loss that a water system operator can determine whether the water loss is a real loss or an unavoidable loss. So, the appropriate data collection from the water meters is the primary important factor in accounting the water use and loss from a water distribution system (Farley and Trow 2003) and the important data needed to assess the water use and loss includes:

1. Information related to the infrastructure like water meters, water mains, water service lines, valves, fire hydrants, customer water meters, storage reservoirs and bulk metering of water imported and water exported of a water distribution system
2. Data related to the quantification of potable water supplied to the water distribution system including both surface and groundwater
3. Quantification of consumed/metered and non revenue water lost with the water distribution system
4. Maintenance and operational activities within a distribution system like:
   (a) Pressure reading of the continuous water system
   (b) Maintenance activities related to number of water main repairs/breaks each year, blow-offs for freezing concerns or water quality, water main rehabilitation or replacement programs, water main plugging/flushing/swabbing programs, discharges at pressure relief valves etc.
   (c) Maintenance activities related to fire flow testing, physical inspection, pool filling, temporary water services from tanker/truck filling, hydrants, sewer cleaning, leaks on hydrants, etc.
   (d) Valve maintenance activities related to boundary valve between two different pressure zones, pressure-reducing valves within the water distribution

system, maintenance on valve systems, leaks on valves, check valve maintenance and inspection

(e) Water service and curb box inspection and maintenance (leaks on service
    connections)
(f) Active leak detection programs
(g) Reservoir use (filling/emptying throughout the day, cleaning, leakage, etc.)

Water Loss Management Strategies

There are a number of ways which can be used to manage the loss of water; however,
the decision about the use of a particular program depends upon the area of
occurrence of water loss and on the conditions of local water infrastructure. Installation of meters, detection and repairing of leakages in the distribution system,
maintenance of valve management of water pressure (including surge suppression),
improvement of infrastructure, conservation-oriented pricing designing of standards
for construction methods and pipe material and assessment of night time flow are a
few strategies whose singular or mixed use can be helpful in minimizing the water
loss in a water distribution system.

Floods

A condition of complete or partial inundation of two or more acres of normally dry
land is known as flood. Result of an overflow of inland or tidal waters, flood is
arguably the most widespread weather-related hazard that can virtually occur any-
where in the globe. Often thought of as a result of heavy rainfall, it can occur in a
number of ways that are not directly related to the ongoing weather events, and thus,
a complete description of flooding must include processes that may have little or
nothing to do with the meteorological events (Babbitt and Doland 1949). The origin
of flooding is, therefore, ultimately related to the atmospheric processes creating
precipitation, no matter which particular event causes it. Floods produce damage
through the immense power of moving water and through the deposition of dirt and
debris after the recession of the floodwater. The energy of the moving water goes up
as the square of its speed, i.e., when the speed of water doubles, the energy
associated with it increases four folds. Flooding is typically coupled with water
moving faster than normal, because of the weight of an increased amount of water
upstream, leading to an increase in the pressure gradient that drives the flow that
sweeps away everything that come its way. In most cases, the damage potential of
the flood is magnified by the debris that the water carries, leaving behind scenes of
terrible destruction. Moreover, flood waters typically contain suspended silt and
potentially toxic microorganisms and dissolved chemicals, thus compromising
drinking water supplies, resulting in short-term shortages of potable water, with
the additional long-term costs in restoring drinking water services to the residents of

a flooded area. The mud and debris left after the floodwater recession are costly to clean up and also represent a health hazard. In some situations, floods drive wild animals (including invertebrates of all sorts) from their normal habitats into human habitations near and within the flooded areas, which creates various problems, especially when the animals are venomous or aggressive. Although flooding is largely negative in its impacts on humans, it is also considered as a natural process that shapes the earth, as floodplains along streams and rivers are among the most fertile regions known around the world. Furthermore, the so-called "cradles of civilization" also lie within the flood plains prompting us to believe that humans have been affected both positively and negatively by flooding since historical times, whenever they found themselves in the path of these natural events.

Occurrence

Floods that usually occur after spring rains, heavy thunderstorms, and winter snow thaws, are caused by multiple factors including highly accelerated snowmelt, heavy rainfall, tsunamis, unusual high tides, severe winds over water, or failure of levees, dams, retention ponds, or other structures that retain the water. It is exacerbated by other natural hazards such as wildfires, which reduce the density of vegetation thus interfering with the absorption of rainfall or by the increased areas of impervious strata.

Periodic floods occur in many rivers, forming a surrounding region known as the floodplains. During times of rain, some of the water is retained in ponds or soil, some is absorbed by grass and vegetation, some evaporates, and the rest travels over the land as surface runoff. Floods occur when ponds, lakes, riverbeds, soil, and vegetation lose the water absorption potential and then the water runs off the land in quantities that is carried within channels, stream or retained in man-made reservoirs, lakes, and natural ponds (Dingman 1994).

Types of Flood

Following are a few common types of floods.

1. Flash Floods

   Rapid and significant rise in water level with low or no warning time due to an intense and sudden heavy rainfall event is known as flash flood. Flash floods which are common in areas of steep slopes occur when rainfall rates are so high that the ground becomes unable to absorb the water enough quickly to prevent significant runoff. Such events also occur due to some dam or levee failure. Floods occur so quickly that they destroy structures, down trees, and wash out roads in no time. Although flash floods do not last long or cover large areas their sudden onset and strength gives them the ability to create devastation at a much larger scale.

2. River Flooding

Compared to flash floods, river floods occur on a much slower time scale. Caused by the water runoff collected in rivers and streams and their eventual increase to the level that overflows the banks. River floods occur by heavy rainfalls, snowmelts and ice jams. These floods cover an enormous area and affect the downstream areas even if they do not receive much of the rains. Although river flooding can be predicted, its effects, even over a longer period of time, cause extensive damage to residents living near rivers and streams.

3. Coastal Flooding

Tropical storms and hurricanes which cause the large waves to raise the sea level and create a storm surge along the beaches. This phenomenon of pushing the sea water inland is known as coastal flooding. Furthermore, it occurs by the underwater earthquakes which displace large volumes of water to cause waves called tsunamis to rush inland. On a much smaller scale, extremely high tides, sometimes associated with a full moon also cause some minor coastal flooding events.

4. Urban Flooding

Urban floods are the floods that are mostly caused over the urban areas that have a lesser ability to absorb water even in the events of moderate rainfall events. These can also be caused by coastal, river, or flash flooding. Urbanization increases water runoff as much as 2–6 times over what would occur on natural terrain. These floods cause high economic damages to businesses and homes.

5. Areal Flooding

Similar to urban floods, areal floods are the most common floods that usually occur in standing water in open fields and low-lying areas. Such floods often occur due to heavy rainfall over a larger area in a brief period of time. Additionally, a prolonged period of rainfall can also lead to flooding, often causing dangerous inundation of low lying areas. Along with agricultural losses these floods result in the stagnation of water that serves as breeding ground of insects and disease.

Methods of Flood Control

Different structural and nonstructural measures discussed below can be used for the control of floods.

Structural Measures

Dams

Dams and their associated reservoirs are designed partially or completely to aid in flood control and protection. Many large dams have flood-control reservations in which the level of a reservoir is kept below a certain elevation before the onset of the summer/rainy melt season to allow a certain amount of space in which floodwaters are filled. Further dry dams that purely serve for flood control without any conservation storage are also constructed for this purpose.

Water Gates
The water-gate flood barrier is a rapid response barrier which can be rolled out in a matter of minutes. It is unique in the way that itself deploys using the weight of water to hold it back.

Diversion Canals
Diversion canals are purposely built floodways or canals to which the flood water can be redirected in the events of floods. Examples are the flood channels of river Jehlum in Srinagar, the Red River floodway that protects the city of Winnipeg, Canada and the Manggahan floodway that protects the city of Manila, Philippines.

Self-Closing Flood Barrier
A flood defense system designed to protect the people and property from inland waterway flood, is known as self-closing flood barrier. It is designed to protect the residential areas, industrial areas and other strategic areas as it is constantly ready to be deployed in a flood-like situation. Barrier systems have already been built and installed in Italy, Belgium, the Netherlands, Ireland, Thailand, Vietnam, Russia, the United Kingdom, the United States, and Australia. Millions of documents at the National Archives building in Washington DC are protected by two SCFBs.

River Defense
In many countries, rivers are prone to floods and are often carefully managed by defense like bunds, levees, weirs, and reservoirs to prevent the bursting of their banks. When these defenses fail, emergency measures such as portable inflatable tubes, sandbags, or hydrosacks are used.

Coastal Defenses
Coastal defenses like barrier islands, sea walls and beach nourishment and tide gates are used to address coastal floods. They are placed at the mouth of streams or small rivers, where an estuary begins or where drainage ditches, or tributary streams connect to sloughs. Tide gates close during incoming tides to prevent tidal waters from moving open, and open during outgoing tides to allow waters to drain out via the culvert and into the estuary side of the dike.

Temporary Perimeter Barriers
Temporary perimeter barrier is a system accomplished by containing two parallel tubes within a third outer tube. When filled, this structure forms a non-rolling wall of water that controls 80% of its height in external water depth, with dry ground behind it.

Nonstructural Measures

Flood Plain Management and Zoning

The basic concept of flood plain management is to regulate the land use in the flood plains in order to restrict the damage due to floods, while deriving maximum benefits from the same. This is done by determining the locations and the extent of areas likely to be affected by floods of different magnitudes/frequencies and to develop those areas in such a fashion that the resulting damage is minimum in case floods occur (Dingman 2002). Flood plain zoning, therefore, aims at disseminating information on a wider scale so as to regulate indiscriminate and unplanned development in flood plains and is relevant both for protected as well as unprotected area. It recognizes that the flood plains are essentially the domain of the rivers, and as such all developmental activities in flood plains must be compatible with the flood risk involved (Kundzewicz 1999). Following are the broader steps involved in the implementation of flood plain zoning measures:

1. Demarcation of the areas liable to floods and preparation of a detailed contour plans of such areas on a large scale
2. Fixation of reference river gauges
3. Determination of areas likely to be inundated for different water levels and magnitudes of floods
4. Demarcation of areas liable to flooding by floods of different frequencies like once in 2 years, 10, 20, 50, and 100 years. Similarly, area likely to be affected on account of accumulated rainwater for different frequencies of rainfall like 5, 10, 25, and 50 years
5. Delineation of land use type in flood plains in light of the above points

Flood Proofing

Flood proofing is a combination of the structural changes and emergency actions that help in the immediate relief of the flood-prone population and mitigation of their respective distress. It adopts the techniques of providing raised platforms for man and animals and raising the public utility installations and other facilities above the flood level to make the essential services flood proof.

Flood Warning and Forecasting

It enables forewarning as to when the river is going to use its flood plain, to what extent and for how long. As per the strategy of laying more emphasis on non structural measures, Central Water Commission has established a nation wide flood forecasting and warning system. With reliable and advanced warning/information about impending floods, loss of human lives and loss of moveable properties and human miseries can be reduced to a considerable extent. People, cattle, and other valuable moveable properties can be shifted to safer places.

Droughts

Droughts which usually result in an acute water shortage mostly originate from a deficient precipitation over an extended period of time (like a season or more). It is considered relative to some long-term average condition of balance between evapotranspiration and precipitation in an area and is further related to the number and intensity of rainfall events (effectiveness) and timing (including principal season of delays, occurrence of rains in relation to principal crop growth stages, occurrence in the start of the rainy season). In some regions of the world, droughts are further related to some other climatic factors like high wind, temperature, and low relative humidity which significantly aggravate their severity and make them the most damaging environmental phenomena (Subramanya 2008). In general, drought can be taken as a reduction in environmental moisture status relative to its mean state. Due to the complex nature of droughts variable approaches are used for the identification of droughts and one such theory which makes the drought identification by the assessment of intensity, duration, and location of drought from the study of environmental moisture status dynamics, was developed by the end of the 1960s. As it is very difficult to determine the precise beginning and end of droughts, they are considered as one of the most complexes of all natural hazards. Droughts have enormous impacts in area farming and wildlife, often killing crops, animals, edible plants, and trees (Mosley 2014). Wildfires for which the dying vegetation becomes a primary ignition sources are considered as a secondary hazard of droughts, which further leads to extremely dangerous situations of heat waves. Drought prediction on a timescale of a month to a season is based on regional abnormal precipitation or abnormal ocean surface temperature in the tropics, revealed by model calculations although the general regional causes of droughts are often considered outside the content of these global causes.

Types of Drought

Meteorological, hydrological, agricultural, soil moisture drought, socioeconomic drought, famine and ecological drought are the main types of droughts that have been recognized in India.

Meteorological Drought

A situation which describes a reduction in rainfall for a specific period of some days, months, seasons, or years below the long-term average for a specific time is known as meteorological drought or it can be defined as a situation in which the mean annual rainfall is <75% of the normal rainfall. Indian Meteorological Department has classified the droughts into moderate (in which the deficiency of rainfall is between 25 and 50% of the normal rainfall) and severe droughts (in which the deficiency is more than 50%). So far as metrological droughts are concerned the rainfall effectiveness is more important than the amount of rainfall.

Major Causes of Meteorological Drought

Some major causes of meteorological droughts are:

1. Below-average rainfall and lean monsoon
2. Early withdrawal and late onset of monsoons
3. Prolonged breaks in monsoons
4. Re-establishment of southern branch of jet streams

Hydrological Drought

It is a drought condition that is related to the reduction of water and subsequent occurrence of meteorological drought. It has two types, the surface water drought and groundwater drought whose onset generally occurs by two successive meteorological droughts.

Surface Water Drought

A condition of drying up of surface water bodies like lakes, ponds, rivers, streams, reservoirs, and tanks which occurs by the meteorological drought along with the activities like large-scale deforestation is known as surface water drought.

Groundwater Drought

A condition of falling groundwater level due to excessive pumping of groundwater without compensatory replenishment is known as groundwater drought. Such conditions create more or less irreversible groundwater droughts even in conditions of normal rainfall.

Agricultural Drought

A condition, wherein there is loss of moisture and inadequate rainfall to support a healthy crop growth due to extreme moisture stress and crop wilting is known as agricultural drought. Agricultural drought, which is a relative category depending on the values of plants and soil, is concerned with the impact of meteorological and hydrological drought on crop yield. What could be a drought condition for the cultivation of rice could be a suitable condition for wheat and a condition of excess soil moisture for dry crops like bajra or jowar. Thus, the choice of crops evolves according to variations of climatic and soil conditions.

Soil Moisture Drought

A condition of inadequate soil moisture, particularly in rain-fed areas, which may not support crop growth is known as soil moisture drought.

Socio-economic Drought

Drought that reflects the reduction of availability of food and loss of income on account of crop failures by endangering the food and social security of the people is known as socio-economic drought.

Famine

Large-scale collapsing of food accessibility that can lead to mass starvation without intervention is known as famine.

Ecological Drought

Significant failure in the productivity of natural ecosystems as a consequence of environmental damage is known as ecological drought.

Causes of Drought

Droughts generally occur because of the following causes:

1. Increased gap between precipitation and evaporation with precipitation lying far behind evaporation
2. Overexploitation of scarce water resources
3. Intensive cropping patterns
4. Large-scale deforestation as it leads to exposure of top soil to eroding agents
5. Poor land use and excessive population growth

Climate Change as a Factor of Droughts

It is important to make a distinction between weather and climate, while considering climate change as a factor for droughts. Climate is the behavior of atmosphere over longer periods of time, while weather is the same considered on relatively shorter periods of time. Climate change that occurs over longer periods of time can be seen as a change in the weather patterns as for example, the duration and prevalence of drought in western America has increased due to the warming patterns over the last century. Global climate change further affects a variety of associated factors like precipitation, snow melting, increased evaporation, thus increasing the risk of agricultural, meteorological and hydrological droughts. Much of the Mountain West has experienced declines in spring snowpack, especially since mid-century, and these declines are related to a reduction in precipitation and a shift in timing of snowmelt. Earlier snowmelt, associated with warmer temperatures, leads to water supply being increasingly out of phase with water demands.

Drought Management

Following are a few measures that can be taken at regional and national levels to manage the droughts.

National Level

1. Strengthening of the drought monitoring networks to bridge the gap between the desired and existing meteorological and hydrological monitoring networks
2. Use and improvement of information and communication tools for tackling the multifaceted challenges of droughts
3. Improvement of drought forecasting systems
4. Coordination efforts of different departments and ministries to tackle the drought
5. Developing mechanisms for context-based and need-specific forecasting including local language for better understanding

Regional Level

1. Improvement of real-time monitoring through point monitoring and training
2. Improvement in analytical and methodological tools for drought analysis
3. Organization of joint training programs to build human capacity for improved resilience towards droughts
4. Collaborative and effective implementation of drought relief programs
5. Strengthening of effective water and commodities supply systems

Technical Strategies to Mitigate Droughts

Some of the technical strategies to mitigate the adversities of droughts are as follows:

1. Creation of surface water storage facilities by building dams and diversion canals
2. Planning for less dependable yield by increasing the availability of water for agricultural purposes
3. Prevention of evaporatory losses from reservoirs by the application of a layer of chemicals like acetyl stearyl and fatty alcohols
4. Reduction in the conveyance losses by lining of the canal systems to prevent seepage losses
5. Equitable distribution of water supplies by economizing the water usage
6. Maintenance of irrigation system for keeping the system fit and efficient
7. Proper planning of watershed development involving an integrated approach upon hydrologic and physiographic characteristics which include construction of silicon reservation works on crop lands; construction of structures, like Nalla bunding, check dams, contour bunds, percolation tanks, gully plugging, construction of wells and development of rainwater harvesting etc.

References

Adle, C. (2005). *Qanats of Bam: Irrigation system in Bam from prehistory to modern time*. Papers of the National Workshop on Qanats of Bam, UNESCO Tehran Office: Bam.

An, K. G. (2003). Spatial and temporal variabilities of nutrient limitation based on its in-situ experiments of nutrient enrichment bioassay. *Journal of Environmental Science and Health, Part A, 38*, 876–882.

Annan, K. (2002). *World's water problems can be 'catalyst for cooperation'*. General in message on World Water Day. U.N. Press Release SG/SM/8139 OBV/262.

Anon. (1990). *A directory of wetlands in India*. New Delhi: Ministry of Environment and Forests, Government of India.

Anonymous. (1999). Biotechnology and water security in the twenty-first century. Global water scenario and emerging challenges, Madras declaration, 1999. Chennai: M. S. Swaminathan Research Foundation.

Ansari, A. A., Singh, S., & Khan, F. A. (2011). Eutrophication: Threat to aquatic ecosystems. In A. A. Ansari, S. Singh, G. R. Lanza, & W. Rast (Eds.), *Eutrophication: Causes, consequences and control* (pp. 143–170).

Antoniades, D., Michelutti, N., Quinlan, R., Blais, J. M., Bonilla, S., Douglas, M. S. V., Pienitz, R., Smol, J. P., & Vincent, W. F. (2011). Cultural eutrophication, anoxia, and ecosystem recovery in Meretta Lake, High Arctic Canada. *Limnology and Oceanography, 56*(2), 639–650.

Arceivala, S. J., & Asolekar, S. R. (2007). *Wastewater treatment for pollution control and reuse* (Tata treatment for pollution control and reuse). New Delhi: McGraw-Hill.

Babbitt, H. E., & Doland, J. J. (1949). *Water supply engineering*. New Delhi: McGraw-Hill.

Bandela, N. N., Vaidya, D. P., Lomte, V. S., & Shivanikar, S. V. (1999). The distribution pattern of phosphate and nitrogen forms and their inter-relationships in Barul Dam water. *Pollution Research, 18*, 411–414.

Barnaby, W. (2009). Do nations go to war over water? *Nature, 458*, 282–283. https://doi.org/10.1038/458282a.

Bhattacharyya, A., Janardana, S. R., Manisankar, G., & Naika, H. R. (2015). Water resources in India: Its demand, degradation and management. *International Journal of Scientific and Research Publications, 5*(12), 346–356.

Central Ground Water Board (CGWB). (2006). *Dynamic ground water resources of India*. New Delhi.

Chislock, M. F., Doster, E., Zitomer, R. A., & Wilson, A. E. (2013). Eutrophication: Causes, consequences, and controls in aquatic ecosystems. *Nature Education Knowledge, 4*(4), 10.

CWC, Report. (2005). *General guidelines for water audit CWC, report (2005), "General guidelines for water audit and water conservation"*. New Delhi: Ministry of Water Resources.

Das, G. (2000). *Hydrology and soil conservation engineering*. New Delhi: Prentice Hall of India.

Das, S. (2006). Groundwater overexploitation and importance of water management in India – Vision 2025. In *Tenth IGC Foundation lecture*. Roorkee: The Indian Geological Congress.

Dingman, S. L. (1994). *Physical Hydrology*. Englewood Cliffs, NJ: Prentice-Hall.

Dingman, L. (2002). *Physical hydrology*. Upper Saddle River, NJ: Prentice-Hall.

Farley, M., & Trow, S. (2003). *Losses in water distribution networks* (pp. 146–149). London: A practitioner's guide to assessment, monitoring and control, IWA, ISBN1 900222 11 6.

Foster, S., & Mandavkar, Y. (2008). *Groundwater use in Aurangabad – A survey and analysis of social significance and policy implications for a medium-sized Indian City GW MATE case profile collection*. Washington, DC: The World Bank.

Garg, J., Garg, H. K., & Garg, J. (2002). Nutrient loading and its consequences in a lake ecosystem. *Tropical Ecology, 43*, 355–358.

Gleick, P. (1993). Water and conflict. *International Security, 18*(1), 79–112.

Gleick, P. H., & Heberger, M. (2012). Water brief 4. Water conflict chronology. In P. H. Gleick (Ed.), *The world's water* (Vol. 7, pp. 175–214). Washington, DC: Island Press.

Howard, P., & Tchobanogloss, G. (2002). *Environmental engineering*. New Delhi: McGraw Hill.

Hutchinson, G. E. (1957). *A treatise on limnology* (Geography, physics and chemistry) (Vol. 1). New York: Wiley.

Kakade, B. K., & Hegde, N. G. (1998). Sustainability indicators in watershed management. In *Proceeding of the national workshop on watershed approach for managing degraded land in India - Challenges for the 21st century* (pp. 295–302). New Delhi.

Khullar, D. R. (1999). *India: A comprehensive geography*. Punjab: Kalyani Publishers.

Kundzewicz, Z. W. (1999). Flood protection—Sustainability issues. *Hydrological Sciences Journal, 44*, 559–571.

Likens, G. E. (1972). Eutrophication and aquatic ecosystems. In *Nutrients and eutrophication: The limiting nutrient controversy*. Kansas: Allen Press.

Likens, G. E. (2009). *Lake ecosystem ecology: A global perspective* (A derivative of encyclopedia of inland waters). San Diego, CA: Academic Press, US App 2250.

Linsley, R. K., & Franzini, J. B. (1979). *Water resources engineering* (3rd ed.). New Delhi: McGraw Hill.

Liu, C., Wu, G., Yu, D., Wang, D., & Xia, S. (2004). Seasonal changes in height, biomass and biomass allocation of two exotic aquatic plants in a shallow eutrophic lake. *Journal of Fresh Water Ecology, 19*, 4145.

Marsh, G. A., & Fairbridge, R. W. (1999). Lentic and lotic ecosystems. In *Environmental geology* (Encyclopedia of earth science). Dordrecht: Springer.

Mays, L. (2001). *Water resources engineering* (1st ed.). New York: Wiley.

Mir, A. R. (2006). *Plankton dynamics of Wular Lake with emphasis on fish fauna*. Ph D Thesis submitted to Barkatullah University Bhopal, MP.

Mosley, L. M. (2014). Drought impacts on the water quality of freshwater systems; review and integration. *Earth Science Reviews, 140*, 203–214. https://doi.org/10.1016/j.earscirev.2014.11.010.

Mushtaq, B., Raina, R., Yaseen, T., Wanganeo, A., & Yousuf, A. R. (2013). Variations in the physico-chemical properties of Dal Lake, Srinagar, Kashmir. *African Journal of Environmental Science and Technology, 7*(7), 624–633.

Mushtaq, B., Qadri, H., & Yousuf, A. R. (2018). Comparative assessment of limnochemistry of Dal Lake in Kashmir. *Journal of Earth Science and Climatic Change, 9*, 3.

Oweis, T. Y., Prinz, D., & Hachum, A. Y. (2012). *Water harvesting for agriculture in the dry areas*. Boca Raton, FL: CRC Press.

Pandit, A. K., & Yousuf, A. R. (2002). Trophic status of Kashmir Himalayan Lakes as depicted by water chemistry. *Journal of Research and Development, 2*, 1–12.

Paranjape, S., Joy, K. J., Machadeo, T., Varma, A. K., & Swaminathan, S. (1998). *Watershed based development*. New Delhi: Bharat Gyan Vigyan Samithi.

Patel, A. S., & Shah, D. L. (2008). *Water management*. New Delhi: New Age International Publications.

Rani, K. (2016). Environment degradation and its effects. *International Journal of Advanced Education and Research, 1*(7), 92–96.

Russell, A., Persyn, M., Griffin, A., Williams, T., & Wolfe, C. (2014). *The watershed management approach* (Texas Commission on Environmental Quality). Texas: A & M University.

Sachin, S., Manohar, Rambirendra, Hemant, Shyambirendra, & Takar. (2018). A review on sustainable development of rainwater harvesting system in urban and rural areas. *International Research Journal of Engineering and Technology (IRJET), 05*(10), 511–514.

Shah, T. (2005). *Groundwater and human development: Challenges and opportunities in livelihoods creation and environment*. Lecture delivered during the Workshop on Creating Synergy between Groundwater Research and Management in South Asia. Roorkee: National Institute of Hydrology.

Shapiro, J., Lamarra, V., & Lynch, M. (1975). Biomanipulation: An ecosystem approach to lake restoration. In P. L. Brezonik & J. L. Fox (Eds.), *Water quality management through biological control* (pp. 85–96). Gainesville: University Press of Florida.

Subramanya, K. (2008). *Engineering hydrology* (3rd ed., pp. 175–182). New Delhi: McGraw Hill.

Survey of India Map. (2011). *Census of inland waters of Kashmir Valley* (pp. 9–33).

UNEP. (2002). *Vital water graphics: An overview of the state of the world's fresh and marine waters.* http://www.unep.org/vitalwater

United States Department of the Interior Bureau of Reclamation. (1977). *Ground water manual* (pp. 341–349). Washington, DC: United States Government Printing Office.

Vyas, A., Mishra, D. D., Bajapai, A., Dixit, S., & Verma, N. (2006). Environment impact of idol immersion activity in lakes of Bhopal, India. *Asian Journal of Experimental Sciences, 20,* 289–296.

Wang, G. X., Pu, E. M., Zhang, S. Z., Hu, C. H., & Hu, W. P. (1999). The purification ability of aquatic macrophytes for eutrophic lake water in winter. *China Environmental Science, 19,* 106–109.

Wetzel, R. G. (2001). *Limnology* (3rd ed.). London: Academic Press.

Wolf, A., Shira, Y., & Marc, G. (2003). International waters: Identifying basins at risk. *Water Policy, 5*(1), 31–62.

Air Pollution and Its Abatement

2

Abstract

Air pollution, the unusual interference with the quality of atmosphere by way of addition of contaminants such as smoke, dust, smog, chemicals and vapors, unreasonably interferes with the comfortable enjoyment of life and conduct of business. Caused by many sources including both natural and man-made, air pollution results in a series of devastating impacts on man and his environment including the global environmental disasters like climate change, ozone depletion, and photochemical smog formation. Needing a proper and timely attention it requires an effective abatement strategy which controls the quality of air at surface level in residential, commercial, market, industrial, urban, and workplaces. And the effective abatement strategy includes the technical measures to control the gaseous and particulate pollutants, legislative approaches, substitution of raw materials and modification of the processes involved in the day-to-day activities of humans.

Keywords

Air pollution · Air pollution abatement · Air quality assessment · Lapse rates · Electrostatic precipitators

Recently for the first time in the cultural history, man has forced one of the most horrible changes (pollution) into his environment which sometimes in part was pure, virgin, undisturbed, uncontaminated, and quite hospitable. This change, a result of various human activities has brought about an undesirable change in the physical, chemical, and biological characteristics of our environment, harmful to human life, other species, living conditions and their cultural assets. It has further altered the different major and minor components of environment including air, water, and soil.

Air Pollution

According to *US Public Health services* air pollution can be defined as the presence of contaminants such as dust, smoke, smog, or vapors in the outdoor atmosphere in quantities or characteristics and of duration so as to be injurious to human, plant, and animal life and property or which unreasonably interferes with the comfortable enjoyment of life and property (U.S. Public Health Service 1969).

The atmosphere acting as a natural sink for various kinds of pollutants such as CO_2, CO, H_2S, dust particles, oxides of sulfur and nitrogen emitted from different natural source including natural activity like forest fires, volcanic eruption, decay of vegetation, wind and dust storms and anthropogenic activities like industries and automobile exhausts exceed thousand folds the pollutants emitted from natural sources (Sorbjan 2003). If the rate of entry of these pollutants is faster than the rate of their absorption by the atmosphere (Natural Sink) they gradually accumulate in the air and such disturbance in the dynamic equilibrium of the atmosphere by their considerable accumulation affects both the life and the environment.

Types of Air Pollutants Different types of atmospheric pollutants released from a wide variety of sources including natural and man-made can be classified into following types on the basis of their origin, chemical composition, and states of matter.

1. **According to Origin**

(a) *Primary Pollutants*: Pollutants which are emitted directly into atmosphere from various sources, e.g., oxides of carbon, oxides of nitrogen, oxides of sulfur, particulate matter, etc.
(b) *Secondary Pollutants*: Pollutants which are derived from primary pollutants due to the thermochemical or photochemical reactions taking place in the atmosphere, e.g., peroxy acetyl nitrate (PAN) and ozone (O_3), etc.

2. **According to Chemical Composition**

(a) *Organic Pollutants*: These are the compounds containing carbon and hydrogen which may additionally contain elements like oxygen, nitrogen, phosphorus, and sulfur. Hydrocarbons, aldehydes, ketones, and other organic compounds like carboxylic acids, alcohols, ethers, esters, amines, and organic sulfur compounds are also of a greater concern in the field of air pollution.
(b) *Inorganic Pollutants*: These include carbon compounds like carbon monoxide (CO), nitrogen compounds like oxides of nitrogen (NO_X), ammonia (NH_3), sulfur compounds like hydrogen sulfide (H_2S), sulfur dioxide (SO_2), sulfur trioxide (SO_3), and sulfuric acid (H_2SO_4).

3. According to States of Matter

(a) *Gaseous Pollutants*: Pollutants that get mixed with air and do not settle down are called gaseous pollutants. CO, CO_2, SO_X, and NO_X are a few examples of gaseous pollutants.

(b) *Particulate Pollutants*: Pollutants comprising of finely divided solids and liquids often present in colloidal states such as aerosols, smoke, fumes, dust, fog, sprays, and fly ash are known as particulate pollutants.

Air Pollution Sources Sources of air pollution can be broadly classified into two types.

(a) *Natural Sources*: These sources include:
- Volcanic eruptions
- Forest fires
- Natural decay of vegetation
- Wind storms
- Dust storms
- Marsh gases
- Defilation of sand and dust
- Pollen grains

(b) *Man-Made/Anthropogenic Sources*: These sources include:
- Deforestation
- Burning of fossil fuels
- Vehicular emissions
- Rapid industrialization
- Agricultural activities
- Wars

Air Pollution and Meteorology

Meteorology, the study of atmosphere that concentrates on processes of weather and forecasting has a great bearing on air pollution, as it plays an important role in the air pollution control and management. The significance of meteorology in atmospheric pollution was first appreciated when the substantial use of coal for different purposes such as home heating and industrial power led to the incidents of extreme sulfur pollution. In atmosphere the different meteorological parameters have an essential impact on the amount of pollution, e.g., the solar radiation and temperature affects the pollutant quantities as these parameters are responsible for initiating a series of photochemical reactions in the atmosphere. Dispersion, dilution, and transportation of pollutants is also affected by some meteorological parameters like wind speed and velocity, turbulence and stability (Garratt 1992). Similarly, rainfall washes out the particulate matter from the atmosphere and humidity determines the impact of pollutant concentrations on health, vegetation, and property (Kaimal and Finnigan 1994). Further, the meteorological parameters which directly or indirectly influence the three basic parts of the air pollution problem including the sources, the movement of pollutants, and the recipient, interestingly explain the relation between pollution and meteorology.

Wind Direction and Speed

Wind direction and speed is an essential parameter in the prediction of air pollution in an area where the primary pollution sources are high level emitters and are situated in the vicinity of each other in an industrial zone. However, in an area of low level emitters these factors are not important for air pollution prediction. The predictable determination of wind direction is also related to the location of receptors and topographic features which need to be considered in air pollution potential forecasting and site selection for industries (De 1994).

Further, the movement of air both at microscale and mesoscale over a small geographical area is helpful to determine the movement of air pollutants and the difference in wind speed when it moves horizontally, is an essential factor in the estimation of pollutant diffusion from the stacks.

Atmospheric Stability

Atmospheric stability, which is the ability of atmosphere to resist enhanced vertical motions (Dutton 1995), is an essential parameter in meteorology. It has a practical role in accumulation of air pollutants as it greatly influences the dispersion, deposition, and degradation of pollutants. It also regulates the amount to which vertical waves mix the pollution with the upper air. Stability of atmosphere is controlled by vertical temperature alterations in atmosphere and wind speed (Almethen and Aldaithan 2017).

Lapse Rate

The rate at which the atmospheric temperature varies with the variation in altitude is known as lapse rate. It is considered as positive when the temperature declines with the increasing altitude, negative when the temperature increases with the increasing altitude and zero when there is no change in the temperature with the change in altitude. The lapse rate varies diurnally, monthly, and seasonally at any given location.

Environmental Lapse Rate/Ambient Lapse Rate (ELR)

The environmental lapse rate, which varies significantly from day to day, is the actual measure of decrease in temperature with the increasing height above the earth. It is called as the Standard (or average) lapse rate when averaged out for all places and times. Generally it is about 6.5 °C per 1000 m which varies and depends on local air conditions (Miller 2004). It is further influenced by the ground temperature, seasonal variations, and earth's surface features, as for example, the lapse rates are lower over ground than the sea surfaces. The environmental lapse rate can be determined by a vertically moving radiosonde.

Dry Adiabatic Lapse Rate (DALR)

When a dry parcel of air moves upward through the atmosphere, the surrounding air pressure being low takes energy and therefore cools the parcel at a constant rate known as dry adiabatic lapse rate. The rate at which the temperature of rising parcels decreases is calculated as −0.98 °C/100 m (=−9.8 °C/km). Dry adiabatic lapse rate is approximately taken as −1 °C/100 m (=−10 °C/km).

Saturated Adiabatic Lapse Rate or Wet Adiabatic Lapse Rate (SALR)

The wet adiabatic lapse rate takes into account the energy released when water condenses (the latent heat). As the latent heat release is dependent upon temperature and pressure, the wet adiabatic rate is not constant. It means when a rising parcel of air, saturated with water vapors is cooled to the level of dew point, the water starts condensing, thus releasing latent heat and heating up the air parcel. SALR is calculated as −0.44 °C/100 m (=−4.4 °C/km) at an atmospheric temperature of 20 °C (Stern 1976; Critchfield 1987).

Stable Atmosphere

The resistance of atmosphere to enhance the vertical movements of an air parcel is known as stability of atmosphere. When atmosphere is stable, it neither allows the air parcel to move up nor down but keeps it at the same position. However, in the stable phase of atmosphere if at any time an air parcel receives upward force, it comes back to the position where from it was released.

The stable atmosphere or sub-adiabatic conditions occur when environmental lapse rate is less than dry adiabatic lapse rate (Remsberg and Woodbury 1982). Under such conditions, after the air parcel rises upward, it gets cooled adiabatically but at this level the atmosphere surrounding it is comparatively warmer than the air parcel (Fig. 2.1) and causes the rising air parcel to move down to its original position where from it was released (Miller 2004).

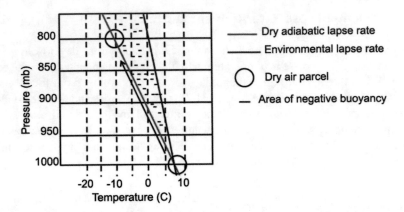

Fig. 2.1 Stable atmosphere

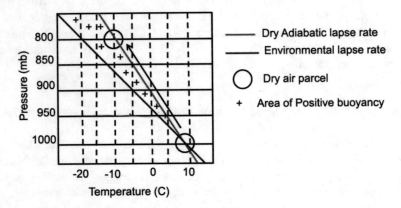

Fig. 2.2 Unstable atmosphere

Unstable Atmosphere

When environmental temperature decreases faster with increasing altitude and the environmental lapse rate becomes greater than dry adiabatic lapse rate, this condition of atmosphere is known as unstable or super adiabatic condition (Miller 2004). Under such conditions when the air parcel rises up, it cools adiabatically but at this level the surrounding atmosphere (Fig. 2.2) is comparatively cooler (Remsberg and Woodbury 1982).

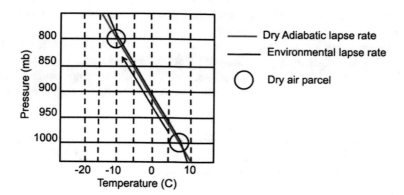

Fig. 2.3 Neutral atmosphere

Neutral Atmosphere

Atmosphere is known to be neutral when both, environmental lapse rate and dry adiabatic lapse rate become equal. Under such conditions if an air parcel moves upward, it cools down adiabatically (Fig. 2.3). The air parcel rises further if it attains some higher temperature after cooling than the surrounding air (Sorbjan 2001). However, the air parcel does not go further upward, if the temperature of the air parcel remains same as that of the surrounding air.

Role of Atmospheric Stability in Dispersion of Pollutants

The pattern and intensity of vertical dispersion of pollutants in the atmosphere is determined by the atmospheric stability, as the same is inhibited by inversions and stable atmospheric conditions, while as, it is favored by unstable and neutral conditions. Existing atmospheric stability and wind speed determine the overall horizontal and vertical dispersion patterns, as the greater the wind speed, the farther the horizontal transport of the pollutants (Sorbjan and Uliasz 1999). Likewise there is less or no dispersion of pollutants if a lower wind speed accompanies the stable atmosphere or inversion. Such conditions favor the continuous build up in the concentration of pollutants in the atmosphere.

Inversion

As a general trend, the higher we go the cooler we feel and it is because at the ground level most of the sun's energy is converted into sensible heat which results in the warming of the air near the surface which on rising up in the atmosphere expands and cools down (Fritz et al. 2008). But sometimes the air temperature actually rises with the rise in altitude and this condition, wherein a warmer air layer covers the cooler air layer on its top is known as temperature inversion. It is known as temperature inversion because here the temperature profile is inverted from its usual state. In such extreme sub-adiabatic condition the vertical movement of air within the inversion layers is almost absent (Stull 1973). The temperature inversion can be divided into major temperature inversion and minor temperature inversion events.

Major Temperature Inversions In stratosphere and thermosphere, the temperature increases with increasing height because most of the ozone of the atmosphere is contained in the stratosphere. As the ozone layer absorbs most of the UV light from the sun it becomes warmer and so does the temperature in this layer with its rising altitude. However, in troposphere the increase in temperature is due to the solar absorptive molecular oxygen.

Minor Temperature Inversions Temperature inversions also occur near the ground or at higher elevations in troposphere, the base layer of the atmosphere. High-level inversions are observed in depressions where the colder air is forced down the warmer air as the cold front undercuts the warm front. However, the low level inversions or ground inversions are generally observed during anticyclonic conditions wherein a rapid heat loss occurs at night due to radiations and such conditions often lead to formation of frost and fog in the valleys.

Minor temperature inversions play an important role in air quality, particularly during the winter seasons when such inversions are strongest. The warmer air acts as a lid above the cooler air, suppresses the vertical mixing, and traps the cooler air at the surface along with the pollutants emitted from different sources like fireplaces, vehicles, and industries near the ground level, thus leading to a poor air quality. The duration and strength of the inversions also controls the ground level air quality index (AQI) as the stronger inversions confine the pollutants to a shallow vertical layer, leading to high AQI levels, while the weaker inversions lead to lower AQI levels.

Maximum Mixing Depths

In lower layers of the atmosphere, dispersion of pollutants is significantly aided by the convective and turbulent mixing. Vertical extent to which the pollutants disperse depends upon the environmental lapse rate which varies diurnally, monthly, and seasonally alongside the topographical features. The depth of this convective mixing layer in which vertical dispersion of pollutants is possible, is known as maximum mixing depth (MMD), e.g., when a parcel of air warmer than the existing ground level temperature rises, it cools according to adiabatic lapse rate and the level where its temperature becomes equal to the temperature of surrounding air gives us the value of maximum mixing depth (Stull 1988). The urban air pollution episodes are generally known to occur when the maximum mixing depth is equal to or less than 1500 m.

The quantity of air available to dilute the atmospheric pollutants is strongly related to the speed of wind and to the degree to which emissions can go up in the atmosphere. The product of the average wind speed within the mixing depth and the maximum mixing depth (sometimes called the mixing height) sometimes acts as an indicator of the dispersive capability of atmosphere known as the ventilation coefficient (m^2/s) and the ventilation coefficient values ≤ 6000 m^2/s are considered as the indicators of high air pollution.

Plumes and Plume Rise

A mixture of gases and particulate matter is known as a plume. Due to their large size and heavy weight the larger particulates possess appreciable settling velocities, and hence settle near the source while as due to their minute size and light weight the smaller particles keep drifting in the atmosphere for a longer time resulting in a dispersion behavior similar to that of gases (Stern 1976). The gaseous nature of the plumes allows pollutants to disperse by simple diffusion, whereby gaseous molecules move randomly from an area of higher concentration to an area of lower concentration.

Stack plume rise which is a conceptual phenomenon depends mainly on two forces, the buoyant force and the momentum force where in the buoyant forces are governed by temperature, heat emission rate of the stack exit gases and the momentum forces are governed by the stack gas velocity. The plume rise which is also dependent on various other factors like wind speed, atmospheric stability, and atmospheric turbulence is an important parameter in the overall air pollution control mechanism as it directly interferes with the dispersion of pollutants over a period of time and space. The more the plume rise, the better is the dispersion of air pollutants, thus resulting in a lower ground level concentration of pollutants (Peavy et al. 1985).

Further plume rise is an important factor in determining the maximum ground-level concentration of pollutants from the sources since it typically increases the effective stack height by a factor of 2–10. As the maximum ground-level concentration of pollutants is roughly related to the inverse square of the effective stack height, the plume rise can reduce the ground-level concentration of the pollutants by a factor of as much as 100 (Suetsugu and Kogiso 2013).

The plume rises to a certain height (Δh) under a given set of conditions which when added to the actual height of the stack gives the effective stack height. The higher the effective stack height the greater is the dispersion rate of the pollutants.

- Building taller stacks and emitting pollutants at higher temperatures can increase the effective stack height.
- Plume rise is also affected by the horizontal wind speed as the higher wind speed bends over and transports the plume more quickly.

Plume Characteristics

Various plume characteristic including temperature of release (buoyancy), rate of release, height of release and environmental characteristics (including wind direction, wind speed, atmospheric stability, and turbulence) influence the dispersion of a plume in the atmosphere (Muralikrishna and Manickam 2017b).

Factors Influencing Plume Rise

The factors that control the behavior of a plume injected into the atmosphere from a stack include:

- Stack engineering
- Meteorological conditions
- Nature of the effluents

Stack design parameters which influence plume rise include the temperature and speed of the effluents, local heat sources creating the convective turbulences, terrain and buildings near the chimney creating the mechanical turbulence, and the number and shape of stacks in the area. Atmospheric turbulence controls both the rate of mixing of the plume with outside air and the motion of the plume before and after mixing. However, the primary factors which determine the intensity and spectrum of turbulence include the terrain roughness, wind speed, and stability. The meteorological parameters influencing the plume rise include the stability and the horizontal wind speed, as denoted by the vertical temperature or potential temperature gradient. Other factors influencing plume rise include the intensity of solar radiations and the type and distribution of cloudiness as even over smooth and uniform terrains, cumulus clouds cause irregular ground heating thus creating large thermal eddies with appreciable vertical components and under such conditions, plumes from even the tall stacks are possibly brought to the ground (Choi et al. 2014).

Plume Behavior

Plume behavior is the dispersal pattern of plume gases in the atmosphere released from the stack sources or it is the geometrical configuration of plumes diffusing with the atmosphere. It depends upon the outlet velocity momentum and buoyancy and rises from the chimney to a certain height at the starting point of plume path. Plume behavior further depends upon the local air conditions along with the vertical temperature gradient and varies from season to season, day to day, and day to night. Following are a few major plume patterns (Muralikrishna and Manickam 2017a).

Fig. 2.4 Fanning plume

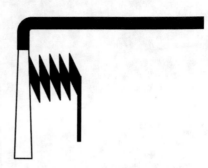

Fig. 2.5 Fumigation plume

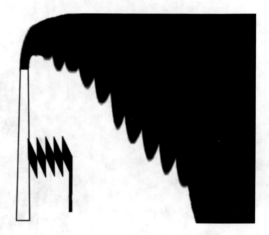

Fanning It is that plume pattern in which the plume has a very little vertical spread and a very large horizontal spread (Fig. 2.4). Such a type of plume typically occurs during nights under a very stable boundary layer with weak variable winds and strong surface inversion as the inversion does not permit the vertical movement of the plume and hence no or low vertical dispersion of the pollutants.

Fumigation It is that plume pattern in which the plume material is brought down rapidly to the ground level due to downward mixing (Fig. 2.5). This type of plume usually occurs shortly after sunrise due to surface heating which is slowly replaced by an unstable layer that grows up to the top of the plume. Such conditions are usually short lived but result in the highest ground level concentration of pollutants.

Looping It is a wavy character that occurs under extremely unstable and convective conditions during afternoon and midday in which large convection eddies are formed (Fig. 2.6). Such large convection eddies take the plume material up and down as it disperses vertically.

Fig. 2.6 Looping plume

Fig. 2.7 Coning plume

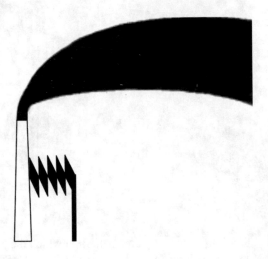

Coning It is that plume pattern in which the plume looks like a cone both in the vertical as well as the horizontal scale and it usually occurs under windy and cloudy conditions (Fig. 2.7). Calculations show that when horizontal wind velocity exceeds 32 km/h and under the condition of cloud blocking of solar radiations during day time and terrestrial radiation during night time, neutral plume tends to form a cone-like structure called as coning plume.

Lofting It is that plume pattern in which the plume stays above the surface inversion and it usually occurs near sunsets when there occurs a transition from unstable to stable conditions (Fig. 2.8). Depending on the rate of deepening of the inversion layer and the stack height, the lofting condition may persist for several hours or may be very transitory in nature.

Fig. 2.8 Lofting plume

Fig. 2.9 Trapping plume

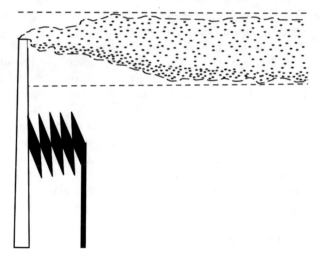

Trapping It is that type of plume pattern in which the plumes disperse their materials uniformly throughout the air (the Planetary Boundary Layer PBL) when released in an unstable atmosphere (Fig. 2.9). When the inversion layer is low and winds are week it can lead to very high ground level concentration of pollutants.

Dispersion of Pollutants

Different accidents and natural events are responsible for emitting toxic and harmful chemicals into the atmosphere, which travel thousands of miles from the point of their release across the globe depending on their physical and chemical properties. They affect the human, animal, and plant health and result in a long-term effect on the rest of the environment (Almethen and Aldaithan 2017).

In troposphere the dispersion of air pollutants is generally governed by advection (wind) field; however, other processes like radioactive decay or turbulent diffusion (turbulence), deposition of air pollutants and chemical reactions play an important role in the spatiotemporal evolution of dispersion patterns. For simulating the dispersion of air pollutants, various modeling approaches which require complex thinking and interaction of researchers from different fields, have been developed. Dispersion of air pollutants is primarily determined by atmospheric conditions, as for example, the super-adiabatic conditions result in a huge vertical air movement and hence enhance the dispersion of pollutants while as the sub-adiabatic conditions produce the opposite characteristics, e.g., during an inversion, vertical air movement becomes almost nonexistent and hence lesser dispersion of pollutants.

During the transport of an air pollutant from a source to a receiver the pollutants disperse into the surrounding air and arrive at a much lower concentration than their concentration on leaving the source. In order to estimate the reduction that occurs during the transport, different atmospheric dispersion models are used. At a given location the concentration of an air pollutant is taken as a function of number of variables, including the distance of the receiver from the source, the rate of emission (amount of the pollutants released at the source), the atmospheric conditions with the vertical temperature characteristics, and the wind direction and speed as the most important conditions. Generally, as the air temperature decreases with the increasing height, it results in an unstable atmosphere and tends to mix pollutants into the upper layers of the atmosphere, thus keeping the pollution load moderate to weak at ground level. However, if the upper air is warmer than the lower air (i.e., if the vertical temperature pattern is inverted), the atmosphere is stable, with calm winds and significantly higher concentrations of pollution (Muralikrishna and Manickam 2017a).

The concentration of pollutants is usually expressed in terms of the total mass of the pollutants in a standard volume of air and the most frequently used measure in metric units for particulate or gaseous pollutants is micrograms per cubic meter of air ($\mu g/m^3$). Parts per million (ppm) is another measure that can be used to express the concentration of gaseous pollutants in the atmosphere.

Dispersion Simulations and Turbulence

Due to the dominance of advection in downwind dispersion the upwind turbulent mixing or crosswind is more efficient than molecular diffusion. Atmospheric dispersion can be regarded as a sum of two main effects including the thermal turbulence caused by buoyancy and the mechanical turbulence caused by wind shear (Dutton 1995). As the thermal turbulence is highly anisotropic, a difficult challenge for turbulence models especially in a convective boundary layer leads to the separated treatment of vertical and horizontal turbulence. However, as the horizontal dispersion is dominated by advection, the vertical mixing of the planetary boundary layer (PBL) is caused by turbulence because of the temperature gradients and the large vertical wind shear together with the moderately low vertical wind speeds (Ahrens 1991) indicating that the vertical turbulence is a key process in atmospheric dispersion simulations and the estimation of its strength requires some sophisticated methods.

Indoor Air Pollution

Degradation of the indoor air quality by the presence of harmful chemicals and other materials is known as indoor air pollution. It may also be defined as the presence of some physicochemical and biological contaminants in the indoor environments not normally present in outdoor air of high quality systems that cause irritation or health effects (Saravanan 2004). As the contained areas enable potential pollutants to build up more than those present in open spaces, indoor air pollution can be tenfolds harmful and worse that the outdoor air pollution (Kankaria et al. 2014). Statistical data suggests that in the developing and Third World countries the health implications of outdoor air pollution are far outweighed by the health impacts of indoor air pollution. Household air pollution from solid fuels is still ranked third among the different risk factors in the report of Global Bureau of Diseases, although there is a decrease in the same in Southeast Asia (Bruce et al. 2000). Most of the sources of these indoor air pollutants are man-made, although some are also natural; however, it is very difficult to provide a comprehensive list of the sources that would be applicable to all indoor environments. In developing countries, the ambient air pollution is far outweighed by the problem of indoor air pollution and in developed countries the most important indoor air pollutants are radon, asbestos, volatile organic compounds, pesticides, heavy metals, mites, molds, and environmental tobacco smoke.

Causes of Indoor Air Pollution

Majority of households using solid fuels burn them in open fire places or simple stoves resulting in the release of a huge quantity of smoke and particulates, thus making it the primary cause of indoor air pollution problems. Inadequate ventilation further increases the indoor pollutant levels by not carrying indoor air pollutants out of the home and by not bringing in enough outdoor air to dilute emissions from the indoor sources (Patel and Aryan 1997). No home is resistant to the poor indoor air pollution problems resulting from various primary sources including:

1. Indoor air pollutants from biotic origin such as from viruses, bacteria, molds, fungal spores, pollens, and insect droppings.
2. Chemicals such as volatile organic compounds (VOCs) and chemicals resulting from the use of cleaning solutions, disinfectants, air fresheners, insecticides, paints, perfumes, adhesives, laminated furniture, hair sprays, cigarettes, gas, kerosene, and coal.
3. Household equipment such as heating and cooling systems, humidification devices, cooking devices like stoves, furnaces, and space heaters emitting particulates and other noxious chemicals.
4. Different sources of outdoor air pollution such as radon, insecticides, fungicides, other pesticides also significantly contribute to the indoor air pollution problems.
5. Combustion products of unprocessed solid biomass fuels used by the rural and urban poor populations for heating and cooking purposes.
6. Some building materials, the ground under the buildings and the bioaerosols also act as a source of some indoor air pollutants.

Besides these common and primary sources of indoor air pollution the following sources resulting in a number of effects are also the significant contributors to indoor air pollution problems.

Environmental Tobacco Smoke (ETS)

The mixture of smoke that comes out from the burning end of pipes, cigars, cigarettes, and the smoke exhaled by the smokers is known as environmental tobacco smoke (ETS). It is a mixture of over 4000 chemical compounds with more than 40 of them being strong irritants and carcinogenic agents. It is a source of volatile organic compounds (VOCs) including polyaromatic hydrocarbons, aldehydes, ketones, organic bases like nicotine, organic acids, and the respirable particulate matter (RPM) (Desai et al. 2004). Environmental tobacco smoke is often referred to as "second hand smoke" and its exposure as "passive smoking."

Environmental tobacco smoke puts the infants and the young children at a higher risk of lower respiratory tract infections like pneumonia and bronchitis. The children who are exposed to environmental tobacco smoke are more likely to have different symptoms of respiratory irritations like excess phlegm, cough, wheezing, and slightly reduced lung function. The US Environmental Protection Agency (EPA) has estimated that the passive smoking annually causes 150,000–300,000 lower respiratory tract infections in infants and children fewer than 18 months of age. Exposure to secondhand smoke further causes nose, eye, and throat irritation along with its effects on the cardiovascular system, as some studies have linked its exposure to the incidents of onset of chest pain in some people.

Biological Contaminants

The biological contaminants including bacteria, viruses, fungal spores, molds, mildews, house dust mites, animal dander and cat saliva, and pollens originating from multiple places inside a home, as these substances act as strong sources of indoor air pollution. Biological contaminants, produced by living things, are often found in damp or wet areas such as humidifiers, cooling coils, unvented bathrooms, draperies, carpets, beddings, and other such spaces and areas. Contaminated central air handling systems can become the breeding grounds of some mildews, molds, and other sources of biological contaminants and can distribute these contaminants throughout the home.

Biological contaminants induce a number of health problems like watery eyes, sneezing, shortening of breaths, coughing, lethargy, dizziness, fever, and digestive problems in humans. They also trigger certain allergic reactions including pneumonitis, hypersensitivity, allergic rhinitis, and some types of asthma (Saravanan 2004). Furthermore the molds and mildews which release certain disease-causing toxins into the air are also problematic for the human health.

Combustion Gases

There is a long history of association between the indoor air pollution and combustion that dates back to the first human dwellings and the respective use of fire. Combustion pollutants are the common contaminants of both indoor as well as outdoor environments. Combustion of the solid, liquid, or gaseous fuels inside a building contributes to the increased concentration of some stable inorganic gases and volatile organic compounds. With the particulate matter, oxides of sulfur and nitrogen, carbon monoxide, hydrocarbons and other odor-causing chemicals as the most common indoor air pollutants, the quantity of release of these pollutants from a source depends upon the fuel/oxidant ratio, the type of fuel, and other combustion conditions.

Carbon monoxide (CO) as a combustion product is an odorless and colorless gas that has an increased efficiency to combine with the hemoglobin and the subsequent interference with the oxygen delivery to the different body parts (Desai et al. 2004). Its lower doses cause a range of symptoms including headaches, dizziness, nausea, weakness, fatigue, and disorientation; however, its higher doses can even cause unconsciousness and death. Of the oxides of nitrogen, nitrogen dioxide (NO_2) is again an odorless and colorless gas that irritates the mucous membranes in the nose, eyes, throat and causes shortening of breaths after exposure to its high levels. Exposure to its higher levels or continued exposure to its lower levels further increases the risk of lung diseases such as emphysema and some respiratory infections in humans. Due to incomplete combustion of certain fuels huge quantities of particulates released into the indoor environment lodge in the lungs and cause damage and irritation of lung tissues. Pollutants, including radon and benzo(a) pyrene, both acting as carcinogenic agents attach to small particles that are inhaled and carried deep into the lung.

Household Products

Organic chemicals used as the common ingredients of certain household products like varnishes, paints, waxes, cleaning and disinfecting agents, cosmetic and degreasing agents and certain fuels act as strong indoor air pollutants as they get released into the indoor environment while using and storing.

The ability of organic chemicals to cause health effects varies greatly from those that are highly toxic to those with no known health effects. Headache, eye irritation, respiratory tract irritation, visual disorders, memory impairment, and dizziness are among the immediate symptoms that are experienced by some people soon after exposure to such organics. Further many organic compounds are suspected or are known of causing cancer both in humans and animals (Kankaria et al. 2014).

Formaldehyde

The prevalence of formaldehyde contamination as a problem is related to the widespread use of wood products bonded with urea-formaldehyde resins, in the construction of cabinets, apartments, and furniture items as the building surveys have shown that more the quantity of these products installed in a building more is the concentration of formaldehyde in the indoor air. Its contamination is considered as the most common cause of residential health complaints.

Exposure to higher levels (above 0.1 ppm) of formaldehyde, a colorless, pungent-smelling gas, can cause burning sensations in eyes and throat, watery eyes, difficulty in breathing and nausea in people. While as evidence suggests that further high level exposure to formaldehyde can trigger asthma attacks in people. Formaldehyde has also been shown to cause cancer both in humans and animals (Kankaria et al. 2014).

Pesticides

Pesticides which include products used to control insects, termites, rodents, fungi, and microbes and sold as liquids, sprays, powders, sticks, balls, crystals, and foggers act as strong air pollution sources as they are commonly used in and around the homes. However, the other possible sources of the pesticides include the stored pesticide containers, the household surfaces that collect and release the pesticides, and the contaminated dust or soil that floats or is tracked in from outside.

Exposure to high levels of cyclodiene pesticides induces a number of symptoms, including headaches, weakness, dizziness, tingling sensations, muscular twitching, and nausea in people exposed to it.

Asbestos

Based on its fiber strength and heat-resistant nature, asbestos is commonly used in a variety of building construction material as fire retardant and for insulation. It is used in a wide range of manufactured goods like ceiling and floor tiles, roofing shingles, asbestos cement products, and paper products. It is due to the cutting, sanding, or other remodeling activities of asbestos-containing materials that elevated concentrations of airborne asbestos occur in the indoor environments. Inappropriate ways and means to remove these materials also release the fibers of asbestos into the indoor home air, thus increasing asbestos levels and endangering the people living in those environments.

As the most dangerous asbestos fibers are too small to be visible, they are inhaled and accumulated in the lungs. Their accumulation in the lung tissues is known to cause lung cancer, mesothelioma (cancer of the abdominal linings and chest), and asbestosis (irreversible lung scarring that can be fatal).

Radon

Radon, an odorless and colorless gas is the primary source of indoor air pollution. Radon due to its high density, sinks in air and is therefore often found in the basements of homes particularly in areas where a lot of boulders and shale are present in the soil. Most of the radioactivity inside a building is associated with radon, emitted from uranium present in the rocks or soil of which the buildings are built. It is a product of radioactive decay process beginning from uranium-238 and thorium-232 as because of their longer half-life periods (4.5 and 14 billion years, respectively) they are present in trace quantities in a variety of geological materials. Besides this, many dwellings and buildings are constructed right on top of radon-emitting rocks itself. Radon daughters are often attached to dust and are present in nearly all airs. However, the background radon levels in the outdoor air are generally quite low of the magnitude of about 0.003–2.6 picocuries of radon per liter of air. Radon sometimes enters into the home environment through well water. Being a gas, radon escapes from construction materials, penetrates through cracks in buildings, and is released into the indoor atmosphere where it may be inhaled.

Although a very little information is available on whether radon can penetrate the skin or not, but some radon is expected to penetrate the skin while coming in contact with radon-contaminated water during different activities like bathing. Noncancerous diseases such as thickening of certain lung tissues may occur within a few days or weeks of high-level exposure to radon. However, chances of getting lung cancer may occur due to the long-term exposure to radon and its daughters present in air and it takes several years before the actual effects become apparent.

Control Measures of Indoor Air Pollution

Following are a few measures used to control the indoor air pollution (Kankaria et al. 2014):

1. Elimination of individual sources of pollution or reduction of their emissions.
2. Proper sealing and enclosure of the sources which contain asbestos.
3. Increasing the air ventilation rate inside the rooms.
4. No indoor smoking and if indoor smoking becomes unavoidable ventilation in the smoking area should be increased.
5. Reduction of entry of biological contaminants in houses by using high-efficiency central vacuum system filters.
6. Proper adjustment of burners in the cooking stoves and use of exhaust fans in the cooking areas.
7. Minimization of the use and exposure to moth repellents.
8. As an alternative, solid wood products should be used in place of the exterior-grade pressed wood products made with phenol-formaldehyde resin in cabinetry, floor, and furniture items.
9. Installation of an effective moisture barrier prior to installation of carpets on concrete basements in order to prevent the generation of molds.
10. Using the multitasking Sanibulb™ to sanitize, deodorize, and purify the air.

Ambient Air Quality Monitoring

Particulate matter including the respirable suspended particulate matter (RSPM) and suspended particulate matter (SPM), hydrocarbons, oxides of sulfur and nitrogen, carbon monoxide, ozone and some photochemical oxidants are the most frequent pollutants occurring in the urban environments. So the ambient air quality monitoring takes into account the monitoring strategies and mechanisms of checking out the levels of these pollutants.

Criteria Recommendations for the Air Monitoring Stations
Monitoring of ambient air quality is generally carried out to check the compliance of the air quality standards to evaluate the impact of some existing or new sources of air pollution and the respective evaluation of the hazards occurring due to some accidental chemical releases or exposures (Rao and Rao 1996). For the same we need to set up different types of monitoring stations whose site and location depends upon the purpose and use of results of the monitoring programs.

With a minimum requirement of three monitoring stations the location of the monitoring station is dependent upon the wind rose diagram that gives the predominant speed and directions of the winds. As a prerequisite, one monitoring station should be at upstream of the predominant wind direction and other two stations should be at downstream. However, the number of monitoring stations can be increased beyond three depending upon the area of coverage. Frequency of data collection for particulate pollutants is once every 3 days; however, for the gaseous pollutants continuous monitoring is needed.

Ambient Air Monitoring Station Types

Station Type Description

Station Type A
Known as downtown pedestrian exposure station, it is located in the congested areas of central business districts, with an average traffic flow of >10,000 vehicles per day and surrounded by many buildings and pedestrians. Height of the station is 2.5–3.5 m from the ground level and location of station is 0.5 m from the curve (Rao and Rao 1996).

Station Type B
Type B monitoring stations known as downtown neighborhood exposure stations are located in the non-congested areas of central business districts with an average traffic flow of <500 vehicles per day, surrounded by less high rise buildings and the typical locations of these stations are malls, parks, and other landscape areas. Height of the station is again 2.5–3.5 m from the ground level and location of station is 0.5 m from the curve (WHO 1976).

Station Type C
Type C monitoring stations known as residential population exposure stations are located in the midst of the residential or suburban areas. Such stations should be >100 m away from the street.

Station Type D
Type D monitoring stations known as mesoscale stations are located at suitable heights to collect the air quality and meteorological data at upper elevations.

Station Type E
Type E monitoring stations known as nonurban stations are located in remote nonurban areas with no industrial and traffic activity. These stations are installed with the main purpose of monitoring the trend analysis at a location of 0.5 m from curve and a height of 2.5–3.5 m from the ground level.

Station Type F
Type F monitoring stations known as specialized source survey stations are located at specified locations to determine the impacts of an under scrutiny air pollution source on the air quality. Such stations are located 0.5 m from curve at a height of 2.5–3.5 m from the ground level (WHO 1976).

Components of Ambient Air Sampling Systems

Following are the main components of an ambient air sampling system:

(a) *Inlet manifold*: the inlet manifold that should neither be too long nor too twisted transports the sampled pollutants in an unaltered condition from the ambient air to the collection medium or the analytical device.
(b) *Air mover*: These are pumps which provide the required force to create low pressure or vacuum at the end of sampling systems.
(c) *Collection mediums*: These are some solid or liquid sorbents or dissolving gases or filters or chambers required for air analysis.
(d) *Flow measurement device*: The devices such as rotameters used to measure the sampled air volume are known as flow measurement devices.

Characteristics of Ambient Air Sampling Systems

Sample stability, collection efficiency, recovery, minimal interference, and understanding the mechanism of collection are a few main characteristics of the ambient air sampling systems. Of these characteristics the first three should be 100% efficient, e.g., the sorbent material for sulfur dioxide (SO_2) should be such that it removes its 100% from the ambient atmosphere at ambient temperature.

Basic Sampling Considerations

Following are some basic considerations for the sampling purposes:

(a) For accurate analysis the sample should be enough larger.
(b) Sample must be a true representative in terms of time, condition, and location to be studied.
(c) Rate of sampling should be fixed so that it provides for maximum collection efficiency.
(d) Sampling durations must accurately reflect the hourly fluctuations in the level of pollution.
(e) Preference must be given to continuous sampling.
(f) No alteration or modification of pollutants must be done during the collection process.

Sampling Errors by High Volume Sampler (HVS)

Following are the two main possible errors by high volume samplers:

1. Due to long and twisted inlet manifolds the particulate pollutants are lost during sampling.
2. For particulate pollutants, biased results are obtained in absence of isokinetic conditions.

Advantages of High Volume Sampler

Following are the major advantages of a high volume sampler:

- Its flow rate is high at low pressure drop.
- Its storage capacity for particulate pollutants is higher.
- Its collection efficiency is higher and cost lower.
- It results in the collection of particulates as fine as 0.3 μm with a filtration efficiency of 99%.
- There is no moisture regain.
- There is no substantial increase in the air flow resistance.

Stack Gas Emissions

A flue gas stack such as a vertical pipe, channel, or similar structure is a types of chimney through which the flue gases (combustion product gases) are exhausted to the outdoor air. Flue gases which are produced by the burning of fossil fuels (coal, oil, and natural gas), wood or other such fuels in steam generating boilers, industrial furnaces or other huge combustion devices are generally composed of carbon dioxide (CO_2), water vapors and a small percentage of various other pollutants like carbon monoxide, particulate matter, and oxides of nitrogen and sulfur. The flue gas stacks must quite often be tall up to 400 m or more, so as to disperse the flue gas emissions over a huge area thereby reducing the concentration of pollutants to the levels required by the standard environmental regulations and policies.

Stack Height

The height of the stack/chimney of a pollution source like an industry or an industrial complex is termed as stack height. It is the physical provision by which the pollutants are released or emitted into the outdoor air. Well before the tall stacks/ chimneys created the problem of acidic rains over the downwind localities in Western Europe and North America it was thought that the greater the stack height the farther the pollutants are transported and hence diluted before reaching the ground level. However, the observations made it clear that while transporting the pollutants over larger distances by the taller stacks these pollutants underwent certain chemical transformations to form acidic rains in places/countries located downwind. Stack height is considered as an important parameter in calculating the downwind

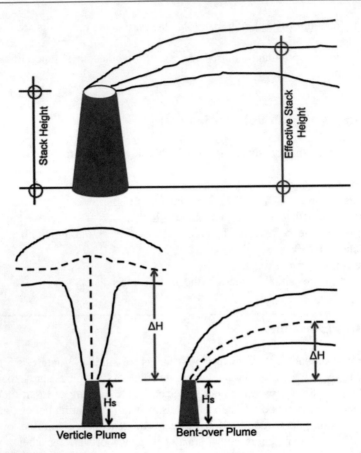

Fig. 2.10 Plume rise and effective stack height

concentration of the pollutants released from different pollution sources as the short stacks transport the pollutants to extremely shorter distances than the tall stacks. However, while considering the stack height, it should be clearly understood that a stack has two different stack heights, one as the physical stack height and the other as the effective stack height. Physical stack height denoted by "h" is the height of the stack from the ground level to the stack top, while the effective stack height symbolized by "H" is the physical stack height plus the plume rise, which is the height upto which the plume raises above the actual top end of the stack (Critchfield 1987). Due to the forced ejection and higher temperature than the surrounding air the plume emitted from the stack takes an upward thrust before taking a bend due to prevailing wind and horizontal transportation in the downwind direction (Fig. 2.10).

Stack Gas Quality Standards

Emission standards which can be local, regional, or national are the permissible emission levels for the specific groups of emitters and all the emitter groups are basically bound to follow these standards by emitting no more than the permitted emission levels. The emission standards which are applicable to any selected group of emitters can be based on some air quality standards or can be entirely independent of any such air quality standards.

These standards prescribe limits of contaminant discharge into the surrounding atmosphere, so that when the standards are met, adverse effects from air pollution are either eliminated or at least minimized. Once air quality standards are established such standards can be used as the basis for formulating emission standards which represent emission levels not to be exceeded if air quality goals are to be achieved. The emission standards can be separately prescribed for different mobile as well as the stationary sources of pollution. Various air quality standards adopted by Central Pollution Control Board (CPCB), India are given in Table 2.1.

Role of Plants in Air Pollution Abatement

It is not surprising to note that plants absorb both natural and anthropogenic pollutants present in the atmosphere as they use what is in their environment, thus making us to believe that they play an increasingly important role in purifying the environment. The pollutants may be absorbed successfully or may cause the vegetation to die after absorption. Researchers at the National Center for Atmospheric Research (NCAR) in Boulder, Colorado, used observations, computer modeling strategies, and gene expression studies to explain how the deciduous plants absorb about a third more of a common class of air-polluting chemicals than previously thought. Thomas Karl, a scientist at NCAR, said that plants clean our atmosphere to a greater extent than we have realized, as they actively consume certain type of air pollution. Trees are thus an important and cost-effective solution for improving the air quality and reduction of pollution. Trees further help to lower down the air temperatures and hence the urban heat island effect.

Leaves after taking in the pollutants through the stomatal pores get them absorbed by the water present inside their tissues. Some plant species are more susceptible to the uptake of pollution, which negatively affects the growth of the plant. Ideally, those plants and trees should be selected that take up higher quantities of pollutants and are resistant to their negative effects. A study by Tripathi and Ranjan (2017) across the Chicago region showed that trees removed approximately 17 tons of carbon monoxide (CO), 93 tons of sulfur dioxide (SO_2), 98 tons of nitrogen dioxide (NO_2), and 210 tons of ozone in 1991. Further researchers have observed that the deciduous plants take up the oxygenated volatile organic compounds at an increasingly fast rate (Tripathi and Ranjan 2017).

Table 2.1 Air quality standards adopted by Central Pollution Control Board of India

Pollutant	Time weighted average	Concentration in ambient air industrial, residential, rural, and other areas	Ecologically sensitive area (notified by Central Government)
Sulfur dioxide (SO_2), $\mu g/m^3$	Annual[a]	50	20
	24 h[b]	80	80
Nitrogen dioxide (NO_2), $\mu g/m^3$	Annual[a]	40	30
	24 h[b]	80	80
Particulate matter (size less than 10 μm) or PM_{10} $\mu g/m^3$	Annual[a]	60	60
	24 h[b]	100	100
Particulate matter (size less than 2.5 μm) or $PM_{2.5}$ $\mu g/m^3$	Annual[a]	40	40
	24 h[b]	60	60
Ozone (O_3) $\mu g/m^3$	8 h[a]	100	100
	1 h[b]	180	180
Lead (Pb) $\mu g/m^3$	Annual[a]	0.50	0.50
	24 h[b]	1.0	1.0
Carbon monoxide (CO) mg/m^3	8 h[a]	2	2
	1 h[b]	4	4
Ammonia (NH_3) $\mu g/m^3$	Annual[a]	100	100
	24 h[b]	400	400
Benzene (C_6H_6) $\mu g/m^3$	Annual[a]	5	5
Benzo(a)Pyrene (BaP)-particulate phase only, ng/m^3	Annual[a]	1	1
Arsenic (As), ng/m^3	Annual[a]	6	6
Nickel (Ni), ng/m^3	Annual[a]	20	20

Source: National Ambient Air Quality Standards, Central Pollution Control Board Notification in the Gazette of India, Extraordinary, New Delhi, November 18, 2009
[a]Annual arithmetic mean of minimum 104 measurements in a year at a particular site taken twice a week 24 hourly at uniform intervals
[b]24 hourly or 8 hourly or 1 hourly monitored values, as applicable, shall be complied with 98% of the time; they may exceed the limits but not on two consecutive days of monitoring

Air Pollution Control

Following two general approaches can be applied for the control of air pollution:

1. *Confining or controlling the pollutants at sources*: Pollutants can be confined or controlled at source either by bringing such modifications in the process that the pollutants are not formed beyond the permissible limits or by reducing the concentration of pollutants to tolerable limits before their release into the atmosphere.

2. *Dilution of pollutants*: Dilution is the reduction of pollutant concentration in the atmosphere to permissible levels by using tall stacks, by proper community planning and by controlling the process with due regard to the local meteorological conditions to prevent buildup of dangerous ground level pollutant concentrations within the designated areas.

General Methods of Air Pollution Control

Following are a few general methods of air pollution control:

Zoning

Zoning of industries is done on the basis of type and function of industry. If zoning is done properly, it results in considerable improvement of the health of a community. As a whole it prevents the invasion of undesirable pollutants of industries in and around residential areas. So, toxic, hazardous, harmful gases and odors are prevented from entering or attacking the human life in residential areas. The industries causing nuisance and producing undesirable gases and odors and other toxic products may be located away from towns in spacious lands.

Control at Source

Control of pollution at source can be done by making use of following strategies:

(a) Substitution of raw material
(b) Modification of process
(c) Alteration of equipment

Installation of Controlling Devices and Equipment

Two categories of devices including those used for the reduction of particulate pollutants and gaseous pollutants are often used for air pollution control.

Control of Gaseous Pollutants

The main gases of concern in air pollution control are the oxides of sulfur, carbon and nitrogen, organic and inorganic acidic, gases and hydrocarbons. Major treatment processes currently used for control of these and other gaseous emissions include:

Table 2.2 Liquid
absorbents used for
the removal of gaseous
pollutants

Pollutant	Absorbent
NO_X	Nitric acid, water
HF	Water, sodium hydroxide
H_2S	Ethanol amine, NaOH + Phenol (3:2)
SO_2	Water, alkaline water

1. Absorption
2. Adsorption
3. Combustion
4. Condensation
5. Masking and counter acting

Absorption

In absorption or scrubbing the contaminated effluent gas (absorbate) is brought in contact with a suitable liquid absorbent so that one or more constituents of the effluent gas are treated, removed, or modified by the absorbent. The absorbent either utilizes the chemical or the physical changes in removing pollutants and the overall efficiency of the process depends upon:

• Chemical reactivity of the gaseous pollutants in the liquid phase
• Extent of surface contact between contaminated gas and the liquid absorbent
• Contact time and concentration of absorbing medium

The equipment used in the process of absorption include plate towers, spray towers, packed towers, bubble cap plate towers, and liquid fit scrubber towers. The gas absorption technique is widely used for removing pollutants like NO_X, H_2S, SO_2, SO_3, and fluorides from gaseous effluents (Table 2.2).

Adsorption

In adsorption the effluent gas stream is subjected to pass through an adsorbent (porous solid material) contained in an adsorption bed. The technique of gas absorption is based on the retention of gases on solid adsorbents by some physical or chemical forces. Physical adsorption depends upon the temperature and pressure conditions in the system. It is promoted by increase in pressure and decrease in temperature and it depends upon Van der Waals forces (intermolecular attractive force). However, the chemical adsorption depends upon the reactivity of gases and their bond forming capacity with the surface of adsorbent. The adsorption materials commonly used are solids such as silica gel, activated alumina, and activated charcoal (Table 2.3). When the waste gas stream contains higher concentration of gases like SO_X and NO_X, they can be recovered economically and used for the manufacture of H_2SO_4 and HNO_3, respectively. The pollutants from power plants can be removed by injecting pulverized limestone into the boiler furnace, where the CaO formed reacts with SO_X to form calcium sulfate and calcium sulfide.

Table 2.3 Adsorbents used for the removal of gaseous pollutants

Pollutant	Adsorbent
NO_X	Silica gel, commercial ziolitic
HF	Porous pellets of Na, limestone
H_2S	Iron Oxide
SO_2	Pulverized limestone or dolomite
Organic vapors	Activated carbon
Petroleum	Bauxite

Masking and Counter Acting

This is a control method chiefly used for odors, by which they are suppressed in masks by the addition of some pleasant odor-producing substances, e.g., addition of vanilla flavors or other flavors to the primary clarifiers of a sewage treatment plant masks the odors of H_2S (hydrogen sulfide) and CH_4 (methane). However, the pleasant odor-producing substances should be nontoxic, noncorrosive, or nonallergic.

Combustion

Though it is a major air pollution source, combustion or incineration is also the basis for an important air pollution control process in which the objective is to convert the air contaminants (usually hydrocarbons or CO) to carbon dioxide (CO_2) and water. For efficient combustion to occur, it is important to have the proper combination of four basic elements including oxygen supply, temperature, turbulence, and time. Depending upon the level of contaminant, direct flame combustion, thermal combustion, or catalytic combustion methods can be used to control the pollution of air.

Controlling Particulate Emissions

Following are a few basic mechanisms for removing particulate pollutants from a polluted gas stream:

1. Gravitational settling
2. Centrifugal impaction
3. Inertial impaction
4. Direct interception
5. Diffusion
6. Electrostatic precipitation

Currently available equipment which make use of one or more of the above removal mechanisms fall into following broader categories.

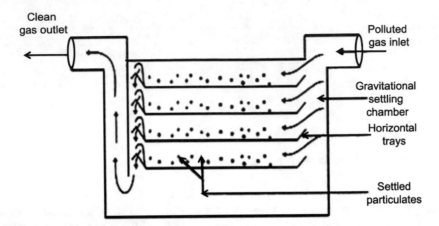

Fig. 2.11 Gravitational Settling Chamber

Gravitational Settling Chambers

These are the devices in which the velocity of horizontal carrier gas is reduced adequately so that particles settle down by the gravitational forces. Particles with diameter ranging between 40 μm and 100 μm are readily collected by this technique (USEPA 1991). The usual collection velocity through settling chambers is between 0.5 and 2.5 m/s; however, for better results the gas flow velocity should be uniformly maintained at <0.3 m/s (Fig. 2.11). The efficiency of gravitational settling chambers is very poor on fine particles and decreases as the load increases.

Advantage:

- Low pressure loss
- Simplicity of design
- Easy maintenance

Disadvantage:

- Requirement of larger space
- Lower removal efficiency for smaller particles
- Removal of larger particles only

Cyclone Separators

A cyclonic collector is a device consisting of a cylindrical shell, conical base, dust hopper, and an inlet where from the dirty air enters tangentially inside the cylindrical shell (Fig. 2.12). In this device the velocity of the incoming gas is transformed into a vortex from which the centrifugal force drives the suspended particulates to the walls

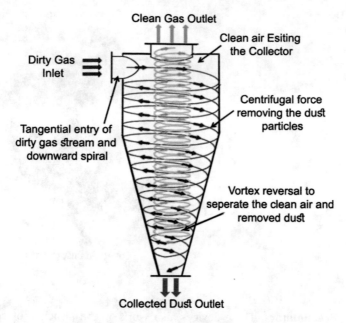

Fig. 2.12 Cyclonic Collector

of the collector. Thus, the sudden change in the direction of gas flow causes the particles to separate out due to their greater momentum. Cyclonic collectors are used to remove particulates from rock product industries, iron industries, steel plants, mining, and metallurgical industries. These devices are best for collection of particles sizing between 15 and 50 μm (Nevers 2000). The efficiency of the cyclonic collectors depends upon the magnitude of the centrifugal force exerted on the particles, as greater the centrifugal force greater is the separating efficiency of the particulates. The magnitude of the centrifugal force generated further depends upon the mass of the particulates, velocity of gas within the cyclone, and the cyclonic diameter.

Advantage:

- Relatively inexpensive
- Simple in design and maintenance
- Requires less floor area
- Low to moderate pressure loss

Disadvantage:

- Requires larger head room
- Lower collection efficiency for smaller particles

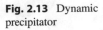

Fig. 2.13 Dynamic
precipitator

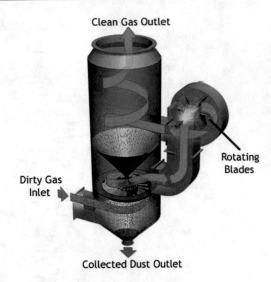

Dynamic Precipitators These devices also work on the principle of centrifugal force. Here the centrifugal force generated by the rotating blades, probes the particles in air stream from where they are drawn off in a concentrated stream (Fig. 2.13). Particles of size between 5 and 20 μm are readily collected by these devices. However, the devices cannot handle wet fibrous materials which get accumulated on the propelling blades.

Wet Collection Devices Devices using mixed phase of gases and liquids are known as wet washers or wet scrubbers. These are the collection devices in which the particles are washed out of the gas flow by a water spray. The most commonly used wet collection devices for the removal of particulate pollutants are:

(a) *Cyclonic Scrubbers*: In such devices water is sprayed at the throat or the entrance of the gas and plates are provided to remove the moisture from the gas after the removal of dust (Fig. 2.14). This is followed by control equipment like gravity setting chambers or cyclonic precipitators. Cyclonic scrubbers can remove dust particles sizing as small as 5 μm with an efficiency of 90% (Nevers 2000).

(b) *Venturi Scrubber*: Here the polluted gas (at a velocity of 60–180 m/s) passes through a duct that has a venturi-shaped throat section (Fig. 2.15). A coarse water spray is injected into the throat, where it is atomized by the high-velocity gas stream and the particulates in the gas stream collide with the liquid droplets to get entrained in the droplets and fall down for removal (Cheremisinoff 2002). Small particles of size between 0.5 and 5 μm associated with smoke and fumes are effectively removed by these highly efficient venturi scrubbers with an efficiency of 90%. Venturi scrubbers can also remove soluble gaseous contaminants.

Fig. 2.14 Cyclonic Scrubber

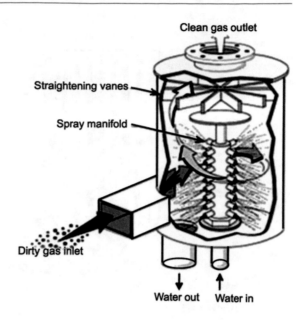

Fig. 2.15 Venturi Scrubber

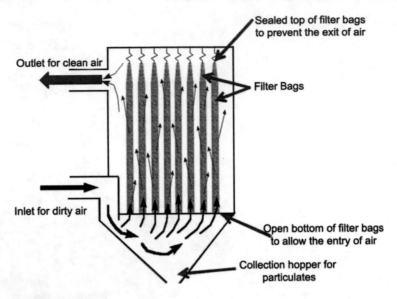

Sealed top of filter bags
to prevent the exit of air

Outlet for clean air

Filter Bags

Inlet for dirty air

Open bottom of filter bags
to allow the entry of air

Collection hopper for
particulates

Fig. 2.16 Bag House Filter

Fabric Filters

In fabric filter systems the dust-laden gases are forced to pass through a porous medium such as woven fabric (Fig. 2.16). The particulates are trapped and collected in the filters and the gases devoid of the particulates are discharged out. Fibrous or deep-bed filters, cloth bag-filters, nylon, dacron, asbestos, silicon-coated glass cloth, etc. are also used as the filter systems. Cloth and nylon filters are used when the temperature is up to 80–90 °C whereas asbestos and silicon coated glass cloth is used at a temperature of 250–350 °C. Wool filters are good for acidic gases whereas cloth, nylon, and asbestos are good for alkaline gases. Fabric filters provide a very efficient method for the removal of particulates even in a size range of <0.5 μm diameter with an efficiency of 99%. The fabric must be cleaned frequently by blowing compressed gas in the reverse direction; otherwise, no gas will be able to pass through it (Schnelle and Charles 2002).

Advantage:

- High efficiency
- Removal of very small particles in dry state

Disadvantage:

- High temperature gasses need to be cooled first
- The flue gasses must be dry to avoid condensation and clogging
- The fabric is liable to chemical attacks

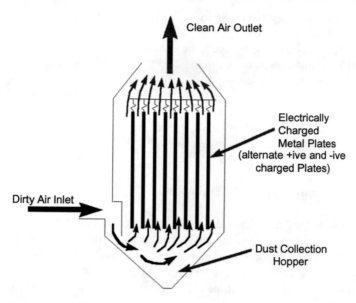

Clean Air Outlet

Electrically
Charged
Metal Plates
(alternate +ive and -ive
charged Plates)

Dirty Air Inlet

Dust Collection
Hopper

Fig. 2.17 Electrostatic precipitator

Electrostatic Precipitators (ESP)

When a gas stream contains aerosols, e.g., dust, fumes, or mist of diameter as small as 0.0001 cm, it can be passed through a more versatile and efficient type of device called electrostatic precipitator. It works on the principle that when the particulates move through a region of high electric potential (30,000–1,00,000 V) they become charged and respectively attracted toward oppositely charged areas where they are collected and removed (Cheremisinoff 2002). The ESP consists of a series of plates which are charged to high voltages alternatively (+) positive and (−) negative. Particles approaching a given plate acquire its charge and are attracted to the surface of next plate from where they fall into the hopper below. Thus particles pick up charge as they pass between plates and are precipitated on plates of opposite charge (Fig. 2.17).

Four basic steps are involved in the operation of an Electrostatic precipitator:

- Electrical charging of particles by ionization
- Transporting the charged particles to the collecting surface
- Neutralizing the electrically charged particles precipitated on the collecting surface
- Removing the precipitated particles from the collecting surface by washing

Electrostatic precipitators prove to be the valuable devices when:

- Very large volumes of gases are to be handled
- Valuable dry material is to be removed
- The gas temperature is very high

Electrostatic precipitators are highly efficient with their efficiency approaching 99.9% and are widely used in power plants, cement industries, paper and pulp industries, iron and steel industries, etc.

Controlling Sulfur Dioxide Emissions

Different methods of sulfur dioxide emission control are based on either the prevention of their emission or end-of-pipe treatment of flue gases containing sulfur dioxide. In sulfur dioxide emission control the substitution of sulfur-containing fuels by clean fuels is a desirable strategy as the small-scale flue gas cleaning is often impractical and hence not feasible. Various approaches for controlling sulfur dioxide emissions include:

- *Use of clean fuel*: It promotes the substitution of high sulfur fuels by low sulfur fuels or else a reduction of fuel-sulfur before its firing. Since sulfur dioxide emissions are directly proportional to the sulfur content of fuel, and also to the amount fired, a reduction in emissions can be achieved by switching to low sulfur fuels and to higher quality ones.
- *Removal of sulfur from the fuel*: It includes washing out of sulfur-rich coal before burning to reduce the content of sulfur that otherwise is emitted by the burning of sulfur-rich coal. Numerous cleaning methods have been adopted for desulfurization of sulfur-rich fuels prior to their firing. However, sometimes these methods are associated with their physical characteristic alterations and hence result in operational problems.
- *Preventing production and release of SO_2 during combustion*: Numerous technologies with the integrated gasification combined cycle (IGCC) and the fluidized bed combustion as the most developed ones have been developed to control sulfur dioxide emission during combustion.
- *Flue gas desulphurization*: The flue gas desulphurization (FGD) technique is basically an end-of-pipe treatment method for the control of sulfur dioxide emissions. Here the flue gas is treated before its release into the surrounding air via the chimney/stack. These systems have been installed on various industrial and utility boilers and on some industrial processes for a number of years and have been observed to efficiently remove the sulfur dioxide emissions from the flue gases. Depending upon the operating conditions of the systems they are capable of removing approximately 70–90% of the sulfur dioxide emissions from the flue gas (Biondo and Marten 1977); however, some systems have achieved a sulfur removal efficiency of >95.2%. All of these techniques can be used

separately, or in conjunction with each other, depending on the quality of the fuel and the emission requirements.

Advantages of Flue Gas Desulphurization (FGD) Technique Following are a few advantages displayed by the FGD technique:

- Use of a clean desulphurized fuel is simpler than installing the huge pollution control equipment at small-, medium-, or large-scale industrial plants.
- Desulphurization helps in attaining an easily recoverable and marketable supply of elemental sulfur from fuel that can be easily shipped and handled.
- A unique feature of flue gas desulphurization is that the gas comes in contact with fresh scrubbing liquid at every stage (Unlike other countercurrent scrubbers) which increases the scrubbing efficiency because of high-mass transfer coefficient.
- Due to open-grid design the flue gas desulphurization has no chocking problems.
- Compared to the conventional scrubbers it has very less downtime thereby reducing maintenance cost.
- Due to its operation in concentrated mode, it has the ability to handle high surges of sulfur dioxide concentrations with less effect on outlet sulfur dioxide levels.

Ozone Depletion

Ozone, a colorless and odorless gas, has a pungent smell, bluish color and is composed of three oxygen atoms (O_3). The name of ozone comes from a Greek word *"ozein"* meaning to smell. The ozone layer was discovered in 1913 by two French physicists Charles Fabry and Henri Buisson. However, a detailed exploration of its properties was done by a British meteorologist G. M. B. Dobson, who developed a simple spectrophotometer (the Dobson meter) that could be used to measure stratospheric ozone from the ground. Regardless of its position of occurrence it has the same chemical structure; however, its useful and harmful nature does depend upon the position of its occurrence in the atmosphere. Naturally the formation of ozone in the upper atmosphere (stratosphere) starts with the dissociation of molecular oxygen into atomic oxygen and such dissociation usually happens by the absorption of sun's radiations with a wavelength of <240 nm. The released oxygen atom combines with the molecular oxygen to form ozone which is being constantly produced and destroyed naturally in the stratospheric belt of the atmosphere at an altitude range of 19–30 km, the reason that 90% of the naturally occurring atmospheric ozone occurs in this layer of the atmosphere (stratosphere). However, normally there occurs a fine balance between the process of ozone formation and ozone depletion thereby safeguarding life on earth. On the basis of its higher concentration in this layer of atmosphere, it is sometimes referred to as "Ozonoshpere" or "Ozone layer" and it filters the ultraviolet radiations coming down from the sun towards the earth.

Formation of Ozone Under natural condition ozone is formed in the stratosphere by the following photochemical reactions:

$$O_2 \rightarrow O + O$$

$$O_2 + O + M \rightarrow O_3$$

where M is the third body which absorbs the excess energy liberated by the above reaction and thereby stabilizes the ozone molecule.

In the lower mesosphere, the atmospheric oxygen dissociates into two oxygen atoms by the absorption of ultraviolet radiations with a wavelength of <240 nm which subsequently react with the molecular oxygen of upper stratosphere to produce ozone. It is also formed at ground level in a very little concentration of about 0.05 ppm where it acts as a harmful pollutant for both plants and animals (Sivasakthivel and Reddy 2011). Nitrogenous gases also react with the ultraviolet radiations from the sun and result in the formation of ozone as follows:

$$NO_2 \rightarrow NO + O$$

$$O_2 + O + M \rightarrow O_3 + M$$

Mechanism of Ozone Depletion The first conscious effort about ozone depletion was made by M. Molino and S. Rowland, two professors of University of California in 1974–1975. The British Antarctic survey team in 1835 and the multinational expedition of Antarctica in 1987 confirmed the predictions about O_3 depletion. They also provided the information about the creation of ozone hole over Antarctica. The problem of ozone depletion and its adverse effects have threatened the existence of life on earth. It is a global problem and is caused by:

1. Natural processes
2. Anthropogenic processes
1. *Natural process*: Under natural conditions the ozone gets depleted by the collision of O_3 with monatomic oxygen in the following manner:

$$O_3 + O \rightarrow O_2 + O_2$$

or

$$O_3 + O \rightarrow 2O_2$$

This natural phenomenon of ozone depletion however does not necessarily upset the ozone equilibrium as it is often compensated by the process of its formation through atmospheric circulation.

2. *Anthropogenic processes*: Anthropogenic processes like industrialization, automobile emissions, nuclear explosions, etc. are also responsible for the depletion

of ozone due to the release of a large quantity of substances like Chlorofluorocarbons (CFCs), Oxides of Nitrogen (NO_X), Hydroxyl radicals (OH) called as ozone depleting substances (ODS). Depletion of O_3 by these substances takes place in the following manner.

By Hydroxyl Radicals (OH)

The hydroxyl radical is generated by either of the two photochemical reactions (2.1) and (2.2):

$$H_2O \rightarrow H + OH \text{ (Hydroxylradical)} \tag{2.1}$$

$$H_2O + O \rightarrow 2OH \tag{2.2}$$

$$OH + O_3 \rightarrow O_2 + HOO \text{ (Peroxide radical)}$$

$$HOO + O \rightarrow OH + O_2$$

$$H + O_3 \rightarrow O_2 + OH$$

By Nitric Oxide (NO)

The nitric oxide is generated by either of the reactions (2.3) and (2.4):

$$N_2O + O \rightarrow 2NO \tag{2.3}$$

$$NO_2 \rightarrow NO + O \tag{2.4}$$

$$NO + O_3 \rightarrow NO_2 + O_2$$

$$NO_2 + O \rightarrow O_2 + NO$$

By Oxides of Nitrogen (NO_X)

Produced by nuclear explosion and introduced directly into the stratosphere:

$$NO_2 + O_3 \rightarrow NO_3 + O_2$$

By Chlorine Radicals

The CFCs used as coolants in refrigerators, air conditioners, as propellants in aerosol sprays and in plastic foams such as "Thermo Cole" or "Styrofoam" are the most important destroyers of ozone layer. The CFC molecules, escaping into the atmosphere, decompose to release chlorine radicals (by photo dissociation), the chlorine radicals released react again and again and hence results in the loss of thousands of ozone molecules. The release of chlorine radicals and the subsequent depletion of ozone by it take place in the following manner:

$$CFCl_3 \rightarrow CFCl_2 + Cl$$

$$CFCl_2 \rightarrow CFCl + Cl$$

$$CFCl \rightarrow CF + Cl$$

$$Cl + O_3 \rightarrow ClO + O_2$$

$$ClO + O \rightarrow Cl + O_2$$

In the above reactions the chlorine atom acts as a catalyst which is not consumed in the reactions and even at the end of the reaction cycle remains as a chlorine atom. Once one atom of chlorine breaks an ozone molecule it becomes free to repeat the process of depletion until it is removed by another reaction in the atmosphere and the basic reason for the same is that chlorofluorocarbon molecules are very stable in nature and can live upto 100 years.

Ozone Depletion Potential (ODP)

Ozone depletion potential, a relative measure with CFC11 taken as a standard reference is the ratio of total amount of ozone destroyed by a particular agent to the total amount of ozone destroyed by the same mass of CFC11 whose ozone depletion potential is usually taken as 1.0 or else it may be defined as the measure of the ability of a compound to destroy the stratospheric ozone. So, if the ozone depletion potential (ODP) of a compound is 0.5 it is roughly half as bad as CFC11. Atmospheric lifetime, molecular mass, number of chlorine or bromine atoms in a molecule and nature of the halogen are a few important factors that affect the ozone depletion potential of any substance.

Effects of Ozone Depletion

As it is an established fact that the ozone layer acts as a protective covering up above the earth by filtering the harmful ultraviolet radiations coming down from the sun. It protects all life forms on earth and any significant decrease in its concentration results in the penetration of ultraviolet radiations and their subsequent reach to the surface of earth leading to diverse harmful effects on all of its organisms (UNEP 1994).

Effects on Climate

Due to weakening of the O_3 layer, there is a less absorption of UV radiations and consequent rise in temperature. This substantial rise in earth's temperature would cause global warming and climate change both at regional and local levels (UNEP 1994).

Effects on Human Beings

(a) Exposures to UV radiations cause skin cancer of three types, viz. basal cell carcinoma, squamous cell carcinoma, and melanoma particularly among the white population.
(b) Ultraviolet radiations damage the langerhans cells in the epidermis of human skin which are the key players in the human immune surveillance.
(c) Damage the eyes—especially leads to the development of cataracts.
(d) By the ultraviolet radiation exposure, the blood vessels near the skin surface are caused to carry more blood thus making the skin hot, swollen, red and hence the sun burns.
(e) Expectedly the ultraviolet radiations are known to cause leukemia and breast cancer although the reasons are not clear.
(f) Suppress the body immune system (responses), thus making the human body more prone to infectious diseases.
(g) Ultraviolet exposure causes mutations in human beings.

Effects on Plants

(a) Ultraviolet radiations cause injuries to plant proteins as they are susceptible to UV injury.
(b) Chlorophyll reduction and harmful mutation are also observed in plants.
(c) Intense Ultraviolet radiation exposure causes greater evaporation.
(d) Typically sensitive plants show reduced growth and smaller leaves.

Effects on Ecosystems

The effects on the aquatic ecosystems mainly depend upon the depth to which the ultraviolet radiations penetrate. Enhanced ultraviolet radiation exposures have shown their serious effect on a range of small organism like zooplanktons, larval crabs, shrimps and juvenile fish as well as slowing rate of photosynthesis in phytoplanktons (UNEP 1994).

Acid Rain

Due to the dissolution of carbon dioxide in the rain water, normally the rain water is slightly acidic. However, due to the presence of different types of pollutants especially the oxides of nitrogen and sulfur (NO_X and SO_X), the pH of the rain water gets further lowered down and this rain with a lowered down pH is known as acid rain or it means the presence of excessive acid in the rain water.

Acid Rain Formation

Vehicular emissions, emissions from thermal power plants, and other types of industries produced by the burning of huge quantities of liquid and solid fossil fuels are responsible for the production of an enormous quantity of oxides of nitrous and sulfur into the atmosphere. In the atmosphere these gases react with the water vapors to form acids which descend on earth as "acid rains" along with the rain water (Kumar 2017). As the polluting gases drift over longer distances in the atmosphere by wind, their presence is felt as far as 2000 km away and the air pollution episodes in one nation can cause acid rains in another nation.

$$SO_X + H_2O \rightarrow H_2SO_4$$

$$NO_X + H_2O \rightarrow HNO_3$$

Effects of Acid Rains

On Human Beings
1. Acid rains have been found to affect the digestive, respiratory, and human nervous systems thus dangerously affecting the life on earth.
2. Acid rains can also cause premature death from lung and heart disorders such as bronchitis and asthma.

On Buildings
1. Acid fumes from the Mathura Oil Refinery have been seen to affect the "Taj Mahal" in Agra as the corrosion due to acid rains results in the formation of crystals of $CuSO_4$ and $MgSO_4$.
2. Acid rains further corrode monuments, statues, bridges, fences and houses.

Global Warming

The unprecedented increase in the earth's atmospheric temperature that actually leads to climate change is known as global warming. Carbon dioxide (CO_2), methane (CH_4), nitrous oxide (N_2O), and chlorofluorocarbons (CFCs) are the four major greenhouse gases which cause adverse effects vis-à-vis the atmospheric temperature and among these carbon dioxide is the most common and important one. Here it is important to note that besides all these gases sulfur dioxide and ozone also act as serious pollutants in causing global warming. These gases known as greenhouse gases cause the atmosphere to trap the increasing amounts of heat energy in the earth's surface to make the planet warmer than usual. Due to a steep incline in the agricultural, industrial, and other human activities the release of more of these greenhouse gases has also seen a steep increase (Agarwal and Narain 2003). Since the beginning of industrial revolution, following human activities have significantly contributed in increasing the atmospheric concentration of these greenhouse gases:

1. Increased use of fossil fuels like coal for power generation and industrial purposes and petroleum for running millions of vehicles
2. Increased rate of deforestation, forest and grassland fires
3. Intensified paddy cultivation by the increased use of inorganic fertilizers resulting in an increased generation of methane and other greenhouse gases

Due to enhanced entrapment of the earth's long wave radiations by the increased amounts of greenhouse gases, the earth's atmosphere is warming up gradually more and more. Since 1960's total atmospheric carbon dioxide has increased from about 320 to over 350 ppm and over this period the average global temperature has also increased slightly by 0.6 ± 0.2 °C. Further the future climate change predictions have indicated that by the middle of the next century, the earth's global temperature may be 1–3 °C higher than what it is today. There is an apparent correlation between increased use of fossil fuels, atmospheric carbon dioxide concentration, and global temperature between 1970 and 2002. Through indirect evidences like coral growth, tree rings, and ice cores, researchers have checked and confirmed that the warmest decade in the past 1000 years was from 1990 to 1999 with 1998 as the warmest year of the millennium followed in order by 2002 and 2001. Observation of the past 33 years of natural disasters by the International Red Cross and Red Crescent has also shown that 90% of the natural disasters were weather related. Moreover, the occurrence of these disasters has increased in the past three decades.

Effects of Global Warming Following are a few major and important effects of global warming:

1. Increased incidence of droughts and heat waves
2. Increased incidences of natural fires in the forest areas
3. Excessive expansion of desert lands
4. Increased rate of evaporation from oceans and other water bodies

5. More cloud formation in the atmosphere
6. Changes in rainfall pattern and disruption of farming activities
7. Rise in sea level due to excessive melting of polar ice caps in Antarctic and Arctic regions
8. Excessive flooding and submergence of low lying coastal areas
9. Longer and hotter summers coupled with shorter and warmer winters
10. Loss of biodiversity due to some direct and indirect impacts on flora and fauna

Control Measures for Global Warming Some of the control and remedial measures for global warming are as follows:

1. Use of alternative resource of energy to reduce the consumption of the increased amounts of fossil fuels like coal and petroleum
2. Increased use of nuclear power plants
3. Increasing the worldwide forest cover
4. Use of unleaded petrol in automobiles
5. Installation of pollution-controlling devices in automobiles and industries

Photochemical Smog

The word smog which is derived from a combination of smoke and fog is the most common example of air pollution that occurs in many major cities around the world. It is a term which refers to air pollution condition in which the atmospheric visibility is partially reduced by a haze generally consisting of smoke and fog (Colbeck and Mackenzie 1994). The smog is generally classified into two types:

(a) *Classical smog*: Due to its association with the traditional fuels like coal this smog is named as classical smog. This type of smog is characterized by higher concentration of sulfur dioxide and unburnt carbon particles. Coal combustion is considered as the main source of this type of smog. Because of the increased presence of sulfur dioxide, a precursor of a weak acid and a mild reducing agent, chemically it has reducing as well as acidic properties and is hence also known as reducing smog or sulfurous smog. It usually occurs in cool and humid climates.

(b) *Photochemical smog*: Characterized by brown hazy fumes the photochemical smog generates by the action of sunlight on unsaturated hydrocarbons and nitrogen oxides produced by petrochemical combustion of automobiles and factories. It generally consists of the oxides of nitrogen, ozone, organic oxidants, aldehydes, hydrocarbons, and other secondary pollutants. Photochemical smog possesses higher concentration of oxidizing agents and is therefore also known as oxidizing smog. It usually occurs in warm, dry, and sunny climates (Colbeck and Mackenzie 1994).

Effects of Photochemical Smog Photochemical smog causes serious health problems along with its impact on rest of the abiotic items:

- Wind ozone and peroxy acetyl nitrate as the principal constituents of photochemical smog it act as a powerful eye irritant.
- Ozone and nitric oxide irritates the throat and nose, and their higher concentration causes headache, dryness of the throat, cough, chest pain, and difficulty in breathing.
- Photochemical smog leads to extensive damage to plant life.
- It also causes corrosion of metals, building materials, stones, painted surfaces, and cracking of rubber.

Control of Smog Formation Following are a few strategies which can be followed to control the formation of smog:

- Smog formation can be controlled by controlling the production of the primary precursors of photochemical smog such as the hydrocarbons and nitrogen dioxide.
- Use of catalytic converters which prevent the release of nitrogen oxides and hydrocarbons into the atmosphere.
- Reduction in the use of such fuels which are responsible for the formation of the primary precursors of the photochemical smog.
- Follow up of the standards laid down for different types of air pollutants.
- Use of air pollution control devices to reduce the amount of pollution generated from different air pollution sources.
- Cultivation of certain plant species like *Pinus*, *Juniparus*, *Quercus*, *Pyrus* and *Vitis* which metabolize the oxides of nitrogen and therefore prove helpful.

References

Agarwal, A., & Narain, S. (2003). *Global warming in an unequal world a case of environmental colonialism*. New Delhi: Centre for Science and Environment.

Ahrens, C. D. (1991). *Meteorology today: An introduction to weather, climate, and the environment*. St. Paul, MN: West Publishing Company.

Almethen, O. M., & Aldaithan, Z. S. (2017). The state of atmosphere stability and instability effects on air quality. *The International Journal of Engineering and Science (IJES), 6*(4), 74–79.

Biondo, S. J., & Marten, J. C. (1977). A history of flue gas desulphurization systems since 1850. *Journal of the Air Pollution Control Association, 27*(10), 948–961.

Bruce, N. G., Perez-Padilla, R., & Albalak, R. (2000). Indoor air pollution in developing countries: A major environmental and public health challenge. *Bulletin of the World Health Organization, 78*(9), 1078–1092.

Cheremisinoff, N. P. (2002). *Handbook of air pollution prevention and control*. Oxford, UK: Butterworth-Heinemann.

Choi, W., Winer, A. M., & Paulson, S. E. (2014). Factors controlling pollutant plume length downwind of major roadways in nocturnal surface inversions. *Atmospheric Chemistry and Physics, 14,* 6925–6940.

Colbeck, I., & Mackenzie, A. R. (1994). *Air pollution by photochemical oxidants, air quality monographs* (Vol. 1). Amsterdam: Elsevier.

Critchfield, H. J. (1987). *General climatology.* New Delhi: Prentice Hall of India.

De, A. K. (1994). *Environmental chemistry.* New Delhi: New Age International.

Desai, M. A., Mehta, S., & Smith, K. R. (2004). *Indoor smoke from solid fuels: Assessing the environmental burden of disease at national and local levels* (Environmental burden of disease series 4). Geneva: World Health Organization.

Dutton, J. A. (1995). *Dynamics of the atmospheric motion.* New York: Dover Publications.

Fritz, B., Hoffmann, W. C., Lan, Y., Thomson, S., & Huang, Y. (2008). Low-level atmospheric temperature inversions: Characteristics and impacts on aerial applications. *Agricultural Engineering International: The CIGR Journal, X.*

Garratt, J. R. (1992). *The atmospheric boundary layer* (p. 316). Cambridge, UK: Cambridge University Press.

Kaimal, J. C., & Finnigan, J. J. (1994). *Atmospheric boundary layer flows* (p. 287). North Carolina: Oxford University Press.

Kankaria, A., Nongkynrih, B., & Gupta, S. K. (2014). Indoor air pollution in India: Implications on health and its control. *Indian Journal of Community Medicine, 39*(4), 203–208.

Kumar, S. (2017). Acid rain-the major cause of pollution: Its causes, effects. *International Journal of Applied Chemistry, 13*(1), 53–58.

Miller, G. T., Jr. (2004). *Environmental science.* Cole, CA: Thomson Brroks.

Muralikrishna, I. V., & Manickam, V. (2017a). *Air pollution control technologies in environmental management* (Science and engineering for industry) (pp. 337–397).

Muralikrishna, I. V., & Manickam, V. (2017b). Air pollution control technologies. *Environmental Management, 11*(7), 2–8.

Nevers, N. D. (2000). *Air pollution control engineering* (2nd ed.). New York: McGraw Hill.

Patel, T. S., & Aryan, C. V. (1997). Indoor air quality: Problems and perspectives. In P. R. Shukla (Ed.), *Energy strategies and greenhouse gas mitigation* (1st ed., p. 72). New Delhi: Allied Publishers.

Peavy, H. S., Rowe, D. R., & Tchobanoglous, G. (1985). *Environmental engineering.* New York: McGraw Hill.

Rao, M. N., & Rao, H. V. N. (1996). *Air pollution.* New Delhi: McGraw Hill.

Remsberg, E., & Woodbury, G. E. (1982). Stability of the surface layer and its relation to the dispersion of primary pollutant in St. Louis. *Journal of Climate and Applied Meteorology, 22*(2), 244–255.

Saravanan, N. P. (2004). Indoor air pollution: Danger at home. *Resonance, 8,* 6–11.

Schnelle, K. B., & Charles, A. B. (2002). *Air pollution control technology handbook.* Boca Raton, FL: CRC Press.

Sivasakthivel, T., & Reddy, K. K. S. K. (2011). Ozone layer depletion and its effects: A review. *International Journal of Environmental Science and Development, 2*(1), 30–37.

Sorbjan, Z. (2001). An evaluation of local similarity on the top of the mixed layer based on large-eddy simulations. *Boundary-Layer Meteorology, 101,* 183–207.

Sorbjan, Z. (2003). In P. Zannetti (Ed.), *Air-pollution meteorology: Theories, methodologies, computational techniques, and available databases and software* (Fundamentals) (Vol. I). Fremont, CA: The EnviroComp Institute.

Sorbjan, Z., & Uliasz, M. (1999). Large-eddy simulation of air pollution dispersion in the nocturnal cloud-topped atmospheric boundary layer. *Boundary-Layer Meteorology, 91,* 145–157.

Stern, A. C. (1976). *Air pollution.* New York: Academic Press.

Stull, R. B. (1973). Inversion rise model based on penetrative convection. *Journal of the Atmospheric Sciences, 30,* 1092–1099.

Stull, R. B. (1988). *An introduction to boundary layer meteorology* (p. 666). Dordrecht: Kluwer Academic.

Suetsugu, D., & Kogiso, T. (2013). Mantle plumes and hotspots. *Module in Earth Systems and Environmental Sciences, 7,* 5–8.

Tripathi, A., & Ranjan, M. R. (2017). Role of plants in mitigation of air pollution. *International Journal of Scientific Research Engineering & Technology (IJSRET), 6*(11), 1087–1094.

UNEP. (1994). In J. C. van der Leun, X. Tang, & M. Tevini (Eds.), *Environmental effects of ozone depletion: 1994 assessment.* Nairobi: United Nations Environment Programme.

U.S. EPA. (1991). *Handbook: Control technologies for hazardous air pollutants,* EPA/625/6-91/014. Cincinnati, OH.

U.S. Public Health Service. (1969). *Air quality criteria for particulate matter* (pp. 148–176). Washington, DC: Department of Health, Education and Welfare.

World Health Organization. (1976). *Manual on urban air quality management.* Copenhagen: World Health Organization.

Environmental Education and Environmental Impact Assessment

3

Abstract

Environmental education is not a separate branch of science but a lifelong interdisciplinary field of study and a way of implementing the goals of environmental protection. It helps us to inculcate the necessary awareness, skills, attitudes, knowledge, and participatory potential in people so that they adjust their day-to-day activities in such a way that they never clash with the environment. Environmental Impact Assessment helps us to analyze both the positive and negative impacts of any proposed activity and the subjective reduction of their negative impacts with the purpose of identification, examination, assessment, and evaluation of the likely and probable impacts and, thereby, helps to work out remedial action plans to minimize the adverse impacts. It is an important management tool for ensuring the justified use of natural resources during developmental process by focusing on the problems, conflicts, or natural resource constraints that could affect the viability of a project. It also examines the implications of a project that might harm people, their homeland or their livelihoods, or other nearby developments. The chapter combines the considerations of how impacts from human activities can be predicted and assessed with the utility of these tools in decision-making, how environmental, economic and social concerns can be balanced, and the potential of the tool to enhance the "spirit of the age," i.e., sustainable development.

Keywords

Environmental education · Environmental impact assessment · Decision-making · Environmental awareness

© Springer Nature Singapore Pte Ltd. 2020
B. Mushtaq et al., *Environmental Management*,
https://doi.org/10.1007/978-981-15-3813-1_3

Environmental Education: Goals, Objectives, Guiding Principles, and Relevance to Human Welfare

Environmental education is a field of education, meant to enhance the protection and conservation of environment. It is an instrument of reducing the environmental degradation and improving the quality of life (Smyth 2006). According to United Nations Educational Scientific and Cultural Organization (UNESCO), "environmental education is not a separate branch of science but a lifelong interdisciplinary field of study and a way of implementing the goals of environmental protection". It is related to those aspects of human behavior which are directly related to the interaction of humans with the natural as well as built environment and their ability to understand this interaction. As in the recent times man has degraded the environment there is an immediate need to make people aware about this degradation so as to change and improve the quality of environment (UNESCO/UNEP 1977).

Objectives

Following are the main objectives of environmental education:

1. *Awareness:* to help the individuals and social groups to gain knowledge about pollution and environmental degradation
2. *Knowledge:* to help the individuals and social groups to gain knowledge about the environment
3. *Attitude:* to help the individuals and social groups to gain knowledge about a set of values for environmental protection
4. *Capacity building and skill development:* to help the individuals and social groups to gain the required skills for making discrimination in shape, form, touch, sounds, habits, and habitats to develop the ability of drawing unbiased conclusions and inferences
5. *Participation:* to enable the individuals and social groups to participate in environmental decision-making

Aims

The main aims of environmental education as highlighted by UNESCO are as follows:

1. To show the sociopolitical, economic, and ecological interdependence of the modern world
2. To develop a sense of responsibility and unity among the different regions and countries vis-à-vis the environmental protection
3. To forge a new international order that guarantees the improvement and conservation of environment

4. To make people understand the complexity of natural and built environment
5. To acquire required knowledge, attitudes, values, and skills to participate in solving and managing the environmental and social problems

Guiding Principles

Following are a few guiding principles of environmental education:

1. Resource Principles
These principles involve the long-term planning and rationale utilization of resources to achieve sustainable development in its truest sense.

2. Soil Principles
As the maintenance and protection of soil is an essential factor for the survival of human civilization and settlements, these principles explain the ways and mechanisms required to maintain a balance in nature alongside the conservation of soil.

3. Wildlife Protection Principles
Due to the biological, scientific, economic, and esthetic importance of wildlife, these principles explain the close association of human survival and wildlife as the life support system, besides highlighting the importance of nature reserves and other wilderness areas in protecting the wildlife.

4. Environmental Management Principles
These principles explain the importance of sound environmental management, waste management techniques, and the influence of human activities and technological innovations on the natural environment.

5. Other Principles
The principles which explain that the relations between humans and their environment are mediated by their culture, i.e., cultural, historical, and architectural heritage are much in need of protection.

Environmental Organizations, Agencies, and Programs

An organization that comes out of the collective conscience of the people who seek to monitor, analyze, and protect the environment against its degradation and misuse by human forces is known as environmental organizations. The organization may be a trust, a charity, a governmental or nongovernmental organization that can be local, regional or national or global in scale (Lieberman 2013). Sierra club, founded on May 28, 1892 in San Francisco, California was one of the first large-scale environmental organization in the world.

List of Some International Organizations

- Earth System Governance Project (ESGP)
- Global Environment Facility (GEF)
- Global Green Growth Institute (GGGI)
- Intergovernmental Panel on Climate Change (IPCC)
- World Wide Fund for Nature (WWF)
- International Union for Conservation of Nature & Natural resources (IUCN)
- United Nations Environmental Programme (UNEP)
- World Health Organization (WHO)
- World Nature Organization (WNO) India
- Agency for Non-conventional Energy and Rural Technology (ANERT)
- Ashoka Trust for Research in Ecology and the Environment (ATREE)
- Babul Films Society (BFS)
- Centre for Science and Environment (CSE)
- Delhi Greens (NGO)
- Environmentalist Foundation of India
- Himalayan Welfare Organization, Pahalgam
- People Cause Foundation
- Save Aravali Trust
- Thakur Mregendra Singh Organization
- The Energy and Resource Institute (TERI)
- Vindhyan Ecology and Natural History Foundation

A number of national and international agencies, organizations, and programs are involved in the different areas of environmental protection and of them some important ones are as follows.

International Bodies

1. Earthscan
Earthscan is an international organization founded by UNEP in 1976. It is working on the commissioning of original articles on matters of environmental concerns and sells the same to magazines and newspapers, specifically in the developing nations.

2. CITES
Convention on international trade of endangered species of flora and fauna is an international form that works in multiple countries on the conservations of endangered flora and fauna by banning their trade. Its membership is open to all the countries and in India, who is signatory to the convention, MoEF works as a nodal agency of CITES.

3. EPA

Environmental Protection Agency is an independent federal environmental agency of the government of United States, founded in 1970. It works on the protection of environment alongside the setting of standards for different environmental parameters.

4. European Economic Community (EEC)

It is a community of 12 European nations, with sound political, economic, and legal base. EEC has programs of implementation and framing of coordinated policy for conservation of nature and natural resources and the subsequent improvement of environment.

5. Human Exposure Assessment Location (HEAL)

Taken up as a health-related monitoring program by World Health Organization in cooperation with UNEP, the project works on three major components including monitoring of air quality, water quality, and contamination of foods on a global basis.

6. International Council of Scientific Unions (ICSU)

ICSU is a Paris-based NGO that works on the exchange of scientific information, initiation of programs requiring international scientific cooperation and reporting on sociopolitical matters and responsibilities in scientific community.

7. International Union for Conservation of Natural and Natural Resources (IUCN)

Founded in 1948, as an autonomous body with its headquarters at Morges, Switzerland, this organization initiates and promotes scientifically based conservation measures for nature and its resources. For the achievement of its goals it works in full coordination with the United Nations and other intergovernmental agencies and sister bodies of World Wide Fund for Nature (WWF).

8. South Asia Cooperative Environmental Programme (SACEP)

This has been recently set up for exchange of professional expertise and knowledge on environmental issue among south Asian member countries including Afghanistan, Bhutan, Bangladesh, Iran, India, Pakistan, and Sri Lanka.

9. United Nations Educational, Scientific and Cultural Organization (UNESCO)

An organization founded in 1945 with its headquarter in Paris to support and implement the efforts of member states to promote scientific research, education, and information, and to provide technical support by organizing seminars and conferences. It supports such activities independently as well as in collaboration with other agencies like UNEP. It further supports activities related to human settlements, environmental quality, training to environmental engineers, and other cultural programs related to environment.

10. World Commission on Environment and Development (WCED)

Set up in 1984 in pursuance to the UNGA (United Nations General Assembly) resolution of 1983, it is a 23 member commission established to reexamine the critical developmental and environmental issues and to formulate proposals for them. It is a call for political action to manage the environmental resources for ensuring human progress and survival. The commission makes an assessment of the level of commitment and understanding of voluntary individuals, organizations, and governmental bodies on environmental issues.

11. Earthwatch Program

Based on a series of monitoring stations and working in coordination with UNEP, Earthwatch monitors the trends in the environment. It was established in 1972 under the terms of the decelerations on human environment.

12. Project Earth

Developed in collaboration with UNEP to inspire interest and educate young people worldwide on the crucial issues facing the Earth's environment. The project is led by Mr. Robert Swan, the UNEP Goodwill Ambassador for youth who is the only person to have reached the North Pole and South Pole on foot.

13. Earthwalk

A series of expeditions designed to focus international attention on environmental issues in key geographical areas. First such walk was taken by R. Swan and six young people were presented by him on June 6, 1992 at United Nations Conference on Environment & Development (UNCED), held at Rio de Janeiro (Brazil).

14. Man and Biosphere Programme (MAB)

This program is an outcome of International Biological Programme (IBP) that has already concluded its activities. MAB was formerly launched by UNESCO in 1971 with 14 project areas under its ambit.

15. World Wide Fund for Nature

It is an international NGO, founded in 1960, that works on the reduction of human footprints on environment as well as in the area of wilderness preservation. It was formerly named as World Wildlife Fund, which remains its official name in Canada and the United States. WWF publishes its living planet report based on ecological footprint and living planet index calculations every 2 years since 1998.

With over 5 million supporters worldwide, working in 100 countries and supporting 1300 conservation and environmental projects, it is the world's largest conservation organization. The group's mission is to stop the degradation of the planet's natural environment and to build a future in which humans live in harmony with nature. Currently, much of its work concentrates on the conservation of three biomes including oceans and coasts, forests, and freshwater ecosystems which contain most of the world's biodiversity. Besides these, it is also concerned with endangered species, sustainable production of commodities and climate change (Malone 1999).

16. United Nations Environment Programme (UNEP)

The United Nations Conference on Human Environment (UNCHE) which was convened in June 1972 lead to the foundation of UNEP (United Nations Environmental Programme), with its headquarter in Nairobi, Kenya responsible for coordination of intergovernmental measures for environmental monitoring. The United Nations Environmental Programme is an agency of United Nations which coordinates its environmental activities and assists developing countries in implementing environmentally sound policies and practices. It was founded by Maurice Strong, its first director, as a result of the United Nations Conference on the Human Environment (Stockholm Conference) in June 1972. It has an overall responsibility for environmental problems among United Nations agencies; however, the international talks on special issues, such as addressing climate change or combating desertification, are overseen by other organizations of United Nations, like the Bonn-based Secretariat of the United Nations Framework Convention on Climate Change and the United Nations Convention to Combat Desertification.

It has played a significant role in promoting environmental science, developing international environmental conventions, and in illustrating ways that can be implemented in conjunction with policy frameworks. It has aided in the formulation of treaties and guidelines on issues such as the transboundary air pollution, international trade in potentially harmful chemicals, and contamination of international waterways (Petsonk 1990). UNEP has created worldwide awareness on emerging and existing environmental issues which lead to the upholding of different international initiatives like Montreal Protocol (1987), Kyoto Protocol (1997), United Nations Framework Convention on Climate Change (1992), and Convention on International Trade in Endangered Species of Flora and Fauna (1973).

It advocates a concept of environmentally sound development, which has led to the adoption of the concept of sustainable development in the Brundtland Commission Report and the United Nations perspective document for the year 2000 and beyond.

Goals of UNEP: The different far reaching goals of UNEP are:

1. Development of national and international environmental instruments
2. Encouragement of new partnerships in the private sector
3. Assessment of environmental terms and conditions
4. Facilitation of transfer of technology and knowledge
5. Promotion of environmental science and information

Table 3.1 International years organized by UNEP

Year	Celebrated as international year
2000	Of thanksgiving For the Culture of Peace
2001	Of Dialogue among Civilizations Of Volunteers Of Mobilization against Racism, Racial Discrimination, Xenophobia and Related Intolerance
2002	For Cultural Heritage Of Mountains and Ecotourism
2003	Of Kyrgyz Statehood Of Freshwater
2004	To Commemorate the struggle against slavery and its abolition Of Rice
2005	Of microcredit, physics, sport and physical education
2006	Of deserts and desertification
2007	International Polar Year
2008	Of Planet Earth, Languages, Sanitation
2009	Of Reconciliation, Natural Fibres, Human Rights Learning, Astronomy and Of the Gorilla [UNEP and UNESCO]
2010	Of Youth, Biodiversity, Seafarer and For the Rapprochement of Cultures
2011	Of Chemistry, Forests, Youth and for People of African Descent
2012	Of Cooperatives and Sustainable Energy for All
2013	Of Water Cooperation
2014	Of Solidarity with the Palestinian People, Crystallography, Family Farming
2015	Of Light and Light-based Technologies, Soils
2016	Of Pulses
2017	Of Sustainable Tourism for Development
2019	Of Indigenous Languages, Moderation, the Periodic Table of Chemical Elements
2022	Of Artisanal Fisheries and Aquaculture
2024	Of Camelids

Source: United Nations observance (2017) https://www.un.org/en/sections/observances/international-years/index.html

Activities

UNEP performs its activities related to resource efficiency, environmental governance, ecosystem management, disasters and conflicts, harmful substances and climate change including the terrestrial approach to climate change (TACC).

International Years Organized by UNEP

The United Nations designates various years as occasions to mark particular events or topics (Table 3.1) in order to promote the objectives of the organization through awareness and action.

17. International Union for Conservation of Nature (IUCN)

Officially known as International Union for Conservation of Nature and Natural Resources, it is an international organization working in the field of sustainable use of natural resources and nature conservation. It was established in 1948 with its headquarter in Switzerland and was previously known as World Conservation Union (1990–2008) and International Union for Protection of Nature (1948–1956). This organization is involved in the gathering of data, analysis of data, education, research and field projects with a mission to encourage, influence, and assist the societies at the international level to conserve nature and to ensure that any use of natural resources is ecologically sustainable. Over the last few decades it has widened its focus beyond the conservation ecology and is now working on the sustainable development-related issues as well. Unlike many other international environmental organizations, IUCN does not itself aim to mobilize the public in support of nature conservation; however, it tries to influence the actions of businesses, governments, and other stakeholders by providing advice and information, and through building partnerships. IUCN is best known for the initiative of compiling the IUCN Red List of threatened species of flora and fauna to assess the conservation status of the species worldwide. It has a membership of over 1400 NGOs and governmental organizations and a voluntary full-time staff of some 16,000 scientists and experts (Petsonk 1990).

Current Work
IUCN Programme 2017–2020
IUCN works on the themes of climate change, business, ecosystems, economics, forest conservation, environmental laws, gender, global policy, marine and polar protected areas, science and knowledge, social policy, species, water and world heritage. Determined by its members, IUCN works on the basis of four-year programs, and in its recent program for 2017–2020 conserving biodiversity and nature is inextricably linked to poverty reduction and sustainable development. This program has identified the following three priority areas for its work:

1. Valuing and conserving nature
2. Supporting and promoting equitable and effective governance of natural resources
3. Deploying nature-based solutions to address societal challenges including food security, climate change, and socioeconomic development

18. World Health Organization (WHO)
World Health Organization was established in 1946 under the United Nations program of attaining highest possible level of health for everyone by acting as the coordinating and directing authority on international health works to establish and maintain objective collaboration with the United Nations, governmental health administrations, specialized agencies, and professional groups. Although it has played a lead role in the eradication of smallpox, its current priorities include communicable diseases with special reference to HIV/AIDS, ebola, malaria and tuberculosis, mitigation of the effects of noncommunicable diseases, sexual and reproductive health, development, ageing, nutrition, food security and healthy

eating, occupational health, substance abuse, and the development of reporting, publications, and networking.

Objectives

To affectively achieve its end results, WHO works on the following objectives:

1. To act as a coordinating authority on international health works
2. To assist governments in strengthening the health services
3. To establish administrative and technical services
4. To work for the eradication of epidemics
5. To promote cooperation among professionals and scientific groups
6. To hold conventions and make recommendations on various health-related matters
7. To conduct research in the field of health
8. To promote mother–child health care to foster the ability and to live harmoniously in the changing environments
9. To promote policies for maintaining environmental hygiene
10. To take all steps to attain the objectives of the organization

Chipko Movement

This is the most famous and powerful people's movement which started in the Himalayan Garhwal region of Uttarakhand. It originally started as the tribal women's protest against felling of trees by contractors but gained momentum under the leadership of Chandi Prasad Bhatt and Sunder Lal Bahuguna. It coincided with the UN Conference on Human Environment held at Stockholm (1972) which recognized the movement as a mighty Environmental Protection Movement. Dasholi Gram Swarajya Mandal spearheaded the movement in Gopeswar, Chamoli district of Uttarakhand. The DGSM under the leadership of Bhatt and Bahuguna started grassroot-level movement involving tribal women to protect the hill ecosystem. Reckless destruction of hill forests by timber contractors caused landslides and floods in the valley of Alaknanda and Bhagirathi and was destroying the fragile hill ecosystem. A unique feature of the movement was the active participation of hill women from villages who were the worst sufferers of deforestation as they had to walk 10–15 km every day to collect their fuel for cooking (Anil 1982). Whenever the contractors with their axmen came for cutting trees, the women hugged the trees and protected them from the axe men. The contractors withdrew from the spot and the forest was saved, and thus the entire Himalayan Garhwal region hill forests were protected from further destruction.

In the course of time the movement spread all along the hill region saving all hill forests and greenery and then moved to the south in Karnataka in 1983 where it was named as "Appiko" movement. Soon it gained international recognition and crossed the geographical boundaries to be observed as Chipko Day at New York, USA in April 1983 where a group of school children assembled at Union Square Park hugged a big tree, followed by some adults. Environmentalists from France, Germany, Sweden, Switzerland, etc., came to visit the Chipko camps and hailed the Chipko Movement (Prasad 1980).

Narmada Dam

Narmada, the largest west flowing river arising from the Kantaka plateau in Shahdol district of Madhya Pradesh is 1300 km in length, draining 9.88 million hectares between the Satpura and Vindhya ranges. Because of inter-water disputes between Madhya Pradesh and Gujarat, the average annual flow of a billion cubic meters of this vast basin is mostly untapped. The government of MP took a major initiative for the development of Narmada basin by involving the construction of some 31 large- and 45 medium-sized dam projects on the Narmada and its tributaries involving a cost of about 25,000 crores rupees. Although the project was aimed to provide several benefits including irrigation to several million hectares of agricultural land, generation of thousands of megawatts of electricity and water supply to thousands of industries etc. (Mehta 2010), the environmentalists and other action groups took the huge dam project as a blueprint for disaster. Because this river basin is one of the most densely forested areas in India and the project would imply the displacement of over one million people, submergence of over 50,000 hectares of agri and forest land and some 1000 villages in the region. In view of these facts and figures it was seen that the environmental damage of the project far outweighs the benefits of the project and in response the environmental action groups, led by environmentalist Smt. Medha Patkar, organized sustained movement to stall the projects of Sardar Sarovar and Narmada Sagar dams and succeeded partly.

Role of Individuals in Environmental Education

Education has the power to modify the society as it can present better knowledge to its populace and can stand as a proper solution to solve different problems existing in a community. Therefore, an individual through environmental education can play a better role in saving the environment. Education is regarded as an important instrument for generating proper awareness and adequate knowledge and skills regarding environmental protection (Gruenewald 2004). It is, therefore, felt imperative to develop education about the environment, for the environment, and through the environment to recognize the role of an individual in environmental education in the following ways.

Awareness
When it comes to environmental protection, the primary aim of education is the awareness to the individuals in a society as everyone in a society including adults,

youth, and kids can understand and become aware of the various environmental issues if they get proper education on how natural environment functions, and how human beings should deal with the environment and ecosystems for sustainability. In the present day, a lot of people conduct environmental awareness programs in schools and societies to help people to get the required awareness about the key environmental issues and take actions accordingly.

Knowledgeable Society to Protect the Environment
An individual can play a huge role in saving the environment by imparting knowledge about the protection of environment to the people in diverse communities. An individual has a potential to curb the environmental problems at various stages through teaching environmental education to the communities and to build knowledge within them to know about the environment and motivate the people to take proper actions to solve the potential problems (Stewart 2001).

Promotion of Holistic Approach
Education on environmental safety or effective environmental protection programs promotes a holistic approach among people to make a fair and sustainable use of resources. Effective programs conducted in schools and other places encourage kids and parents to carry environmental education into their home, and to establish strong bonds with nature as the same helps to promote an ecologically sustainable future.

Environmental Education for Sustainable Development

Following steps need to promote sustainable development through environmental education at individual levels:

1. Promotion of a strong knowledge base on the condition and status of environment
2. Establishment of standards, procedures, protocols, criteria, and recommendations on environmental decision-making
3. Providing integrated solutions to socioeconomic and environmental issues
4. Demonstration of a synergistic association between economic and environmental development
5. Promotion of environmentally sound technology for the conservation of cultural and natural heritage
6. Development of management strategies at different levels to minimize the impacts of environmental problems

Voluntary Agencies and Environmentalists

It is argued that the voluntary agencies can be the only true partner and supporters of the Department of Environment in any real process of change, and to carry the issue of environment to the people. The agencies are free from all types of pressures and are in a position to create public opinion against any decision causing environmental degradation. The nongovernmental organizations provide all the dynamism and vigor to the environmental movements. Several environmentalists stress that government must build links with NGOs and enlist their support in creating the awareness that is needed for ensuring a sustainable development. Environmental programs should give priority to those activities, which can reach the masses in simple language (Stapp 1971). The rural NGOs which deal with the life and death issues of people can help government, to make the task of conservation easy. Government must take them as partners, and coordinate with them for educating the rural masses.

Environmental Impact Assessment (EIA)

An assessment of all possible positive and negative impacts of a proposed activity on the socioeconomic and biophysical environment is known as environmental impact assessment. It is the identification, examination, assessment, and evaluation of the likely impacts of a proposed project activity on the environment thereby working out the remedial action plans to minimize the adverse impacts on the environment. It is an effective management tool that ensures the justified use of natural resources during the developmental process. Environmental impact assessment may also be defined as a systematic process to identify, predict, and evaluate the environmental impacts of proposed actions or projects (Glasson et al. 1994). Environmental Assessment (EA) that ensures the considerations of environmental implications prior to the making of a final decision came into existence since 1970s. With flexibility as its main strength, the process acts as a key tool in environmental management (Ogola 2007). By getting integrated into the project planning process it gives sensitivity to the socioeconomic and environmental impacts of a project, thus helping the project managers to compensate any shortcoming in the project planning process. Although practiced in one or the other form before 1970s, it became an essential part of the common lexicon among the environmental stakeholders as well as the private sectors after Stockholm conference.

EIA focuses on conflicts, problems, or natural resource constraints that could affect the viability of a project, besides examining the project implications that affect the livelihood and household of people or other nearby developments. After predicting the problems, it identifies measures to minimize it and outline the ways to improve the project suitability for its proposed environment. As all the human activities have one or the other kind of impact that may be significant or insignificant, positive or negative, the negative and harmful effects are often far more common than the useful ones. Thus there is a pressing need to evaluate the potentialities of a proposed project before it is undertaken. If the impacts of a project are well within the sustainability of environment there is no danger to environment but if they exceed the carrying capacity they produce ecological changes which must be characterized early in the project cycle to take the necessary corrective steps (Glasson et al. 1994). It is an exercise to document the consequences of a proposed project in totality along with measures necessary to keep the environment healthy and clean (Naber 2012). In order to predict environmental impacts of any developmental activity and to provide an opportunity to mitigate the negative impacts and enhance the positive impacts, the environmental impact assessment (EIA) procedure was developed to find ways to avoid them and to enhance the positive effects.

EIA predicts the constraints and conflicts between the proposed project, program or sectoral plan and its environment. It provides a unique opportunity for mitigation measures to be incorporated into the planning process to minimize the problems. It further enables the monitoring programs to assess and establish future impacts and provide data on which managers can take informed decisions to avoid environmental damages.

Objectives of Environmental Impact Assessment

The Institute of Environmental Management and Assessments (IEMAs) Guidelines for environmental impact assessment identify a number of immediate and long-term objectives of the process of environmental impact assessment.

Immediate Objectives The immediate objectives of the environmental impact assessment process are:

1. To improve the environmental design of the proposals
2. To check the environmental acceptability of the proposals compared to the capacity of the site and the receiving environment
3. To ensure the appropriate and efficient utilization of resources
4. To identify appropriate measures for mitigating the potential impacts of the proposals
5. To facilitate informed decision-making, including setting the environmental terms and conditions for implementing the proposals

Long-Term Objectives The long-term objectives of the process of environmental impact assessment are:

1. To avoid irreversible and serious damage to the environment
2. To safeguard valuable natural areas, resources, and ecosystem components
3. To enhance the social aspects of proposals and protect the human health and safety

Approaches of Environmental Impact Assessment

The general principles useful in evaluating the assessment methodologies are:

(a) *Description of Environment:* Wherever required, the survey of baseline environmental information shall be carried out to determine the existing environmental conditions at the proposed site and in all environs likely to be affected by the proposed project. The issues described in EIA shall be investigated to include existing water and sediment quality, air quality, ecology and environment, the cultural heritage and the man-made environment (WBCSD 2005). The type and duration of baseline surveys shall be such that there is adequate information taking account of all natural variations to define the existing condition. This information shall form the basis for predicting and evaluating the impacts of a project so that the study objectives can be met.

(b) *Impact Prediction:* Assessment methodologies shall be relevant to the issue to be addressed and shall be accepted by the recognized national/international organizations. They shall be capable of:

   1. Identification of potentially harmful and beneficial impacts
   2. Identification of vulnerability of habitats, receivers, or resources to changes
   3. Demonstration of the interaction between the project and its environment
   4. Establishment of the cause and effect relationships in the chain of events
   5. Description and prediction of the reasonable case scenario, worst case scenario, or such scenarios as required in the environmental impact assessment study
   6. Prediction of the extent, nature, and magnitude of the anticipated impacts such that a quantitative evaluation can be done

(c) *Impact Evaluation:* Here, the evaluation of anticipated impacts is done in quantitative terms with respect to a set of criteria. The evaluation methodologies shall be capable of addressing:

   1. The existing or projected environmental conditions with and without the project in place
   2. The sum total of the environmental impacts taking into account all relevant existing, committed, and planned projects
   3. The comparison/differentiation of environmental impacts caused by the proposed projects and that caused by other projects
   4. The extent to which a project aggravates or improves the existing or projected environmental conditions
   5. The environmental impacts during different phases of construction and development of the project
   6. The evaluation of seriousness of the residual environmental impacts

(d) *Impact Mitigation:* The impact mitigation methodologies shall try to avoid the impacts on priority and shall be capable of:

   1. Identification and evaluation of mitigation measures to avoid, reduce, or to nullify the impacts
   2. Assessment of the effectiveness of different mitigation measures
   3. Defining the residual environmental impacts

Benefits of Environmental Assessment

Although the improved project design is an acknowledged contribution of environmental assessments, lack of proper techniques and lack of attention to it has resulted in some weaknesses in the environmental assessment. However, the review of current environmental practices observed the following as the major benefits of the process of environmental assessment:

1. Reduction in time and cost of project implementation along with cost-saving modifications in project designs
2. Enhancement of project acceptance
3. Improvement in project performance by reducing the impacts and breach of rules and regulations
4. Reduction in cleanup/treatment costs

The benefits of environmental assessments for the local communities by taking part in the process include:

- A healthier local environment (water resources, forests, recreational, agricultural, and esthetic potential) and a subsequent improvement in human health
- Biodiversity maintenance and reduced resource exploitation
- Increased skills, knowledge, awareness and decreased conflicts over natural resource use (Arts 2008)

Principles of Environmental Impact Assessment

The basic principles which apply to all stages of EIA should be applied in such a balanced way that it ensures the fulfillment of the purpose and follow up of the internationally accepted standards by environmental impact assessment. Following are a few basic principles which are applied to the different stages of EIA:

Purposiveness: The process should help in an appropriate level of environmental protection, community well-being, and proper decision-making.

Rigorousness: The process should employ the appropriate techniques and methodologies to address the problem under investigation.

Practicality: The process should result in the outcome of best practicable information acceptable to the proponents and helpful in solving the problems.

Relevance: The process should be relevant enough to provide necessary information that is reliable and useful to both planning and decision-making.

Operating Principles: The operating principles describe the applicability of the basic principles to the main steps and specific activities of the environmental impact assessment process including scoping, screening, identification of impacts, and assessment of alternatives.

Efficiency: The process should be time and cost efficient in order to reduce the burden on the proponents and participants consistent with meeting accepted requirements and objectives of the EIA process.

Focused: The process should give due consideration to the key issues and significant environmental effects.

Adaptive: Without compromising the integrity of the process, it should be adjusted to the issues, realities, and circumstances of the proposals under review.

Participative: The process should give ample opportunities for the involvement of affected people and other interested parties and incorporate their concerns and inputs in the documentation and decision-making.

Interdisciplinary: The process should ensure the involvement of appropriate techniques of experts from different relevant fields.

Credible: The process should be carried out with rigor, professionalism, objectivity, fairness, balance and impartiality, and be subjected to independent verifications and checks.

Integrated: The process should also focus on the interrelatedness of the biophysical and socioeconomic aspects of the project proposals.

Transparent: The process should be transparent enough to ensure public access to information, identify the factors that are to be taken into account in decision-making, and acknowledge limitations and difficulties.

Systematic: The process should give due consideration to all the relevant information, the proposed alternatives, their impacts and of the measures necessary to monitor and investigate the residual effects.

Evolution of Environmental Impact Assessment

To properly understand the use of environmental impact assessment as an effective decision-making and environmental management tool, it is important to understand how it evolved in due course of time. In the developed world, United States was the first country to assign mandatory status to EIA through its National Environmental Protection Act (NEPA) of 1969 and since then a number of industrialized countries like Japan, the Netherlands, Australia, and Canada have adopted the EIA legislation in 1984, 1981, 1974, and 1973, respectively and implemented the EIA in their respective countries. In July 1985, the European Community (EC) issued a directive, making environmental assessments mandatory for certain categories of projects (Abaza 2000).

In the developing world, Columbia was the first country to institute a system of EIA in 1974. In Asia and the Pacific region, Thailand and the Philippines have long established procedures for EIA. EIA was made mandatory in 1984 in Sri Lanka. The EIA process in Africa is sketchy, although a number of nations including Rwanda, Botswana, and Sudan have some experience of EIA (Table 3.2).

Some bilateral and multilateral agencies have also recognized the value of EIA as a decision-making tool (Canter 1996). The Organization for Economic Cooperation and Development (OECD) issued recommendations on EIA to its constituent States

Table 3.2 Summary of EIA evolution in different countries

Country	Legislation or declaration under which implementation of EIA evolved
Australia	Environmental Protection (Impact of Proposals) Act 1974, Commonwealth of Australia
Bangladesh	No specific EIA legislation; however, there was a declaration that Environmental Impact Assessments should be carried out for all major developmental projects in 1995
China	Environmental Protection Law, 1979
Canada	Federal Environmental Assessment and Review Process Guidelines Order 1984
France	Law on Protection de la Nature, 1978
India	Environmental Protection Act, 1986
Japan	Principles for Implementing EIA by Environmental Agency, 1984
Malaysia	Environmental Quality (Prescribed Activity) (EIA) Order, 1987
New Zealand	Resource Management Act, 1991
Philippines	Presidential Decree (PD) 1151 Philippines Environment Policy, 1975 PD 1586 Establishing the Environmental Impact Statement (EIS), 1978 Rules and Regulations to Implement the EIS System, 1987
Sri Lanka	National Environmental Act 1980, amended in 1986
Thailand	Improvement and Conservation of National Environmental Quality Act 1975, amended in 1978
The Netherlands	EIA Policy, 1986
USA (California)	California Environmental Quality Act (CEQA), 1971
United States	US Environmental Policy Act, 1969
Vietnam	Environmental Protection Law, 1994
Western Australia	Environmental Protection Act, 1986
West Germany	Cabinet Resolution, 1975

in 1974 and 1979 and for the development aid projects in 1986. United Nations Environmental Programme (UNEP) in 1980 provided guidance on EIA of the developmental proposals and supported research on EIA in developing countries. It further set out goals and principles of EIA for a number of countries in 1987 and provided guidance on its basic procedures in 1988.

The World Conservation Strategy pinpointed the need to integrate environmental considerations with development in 1980 and thus EIA became an integral part of World Bank policy in 1987, stating that environmental issues must be addressed as part of the overall economic policy (Johnson and Fricke 2011). In 1989, the World Bank issued the Operational Directive on Environmental Assessment, which was revised and updated in October 1991 (EC 1999). Asian Development Bank in 1990 published guidelines for EIA and the importance of EIA was further echoed in the Brundtland Report and at United Nations Earth Summit (1992).

Evolution of Environmental Impact Assessment in India

EIA was started in India in 1976–1977, when the planning commission asked the department of science and technology to examine the river valley projects from an environmental angle, which was subsequently extended to the projects which require the approval of the public investment board. The Government of India enacted the Environment (Protection) Act in 1986 and to achieve its objectives, one of the decisions taken was to make EIA statutory. After following the legal procedures, a notification was issued on January 27, 1994 and subsequently amended on May 4, 1994; April 10, 1997; and January 27, 2000 making environmental impact assessment statutory for 30 project activities and hence it became a principal piece of legislation governing EIA in India (Canter 1996). Besides this, the Government of India under this act issued a number of notifications, related to environmental impact assessment which are limited to specific geographical areas, and are summarized as:

- Prohibiting location of industries except those related to tourism in a belt of 1 km from high tide mark from the Revdanda Creek up to Devgarh Point (near Shrivardhan) as well as in 1 km belt along the banks of Rajpuri Creek in Murud-Janjira area in the Raigarh district of Maharashtra (January 6, 1989)
- Restricting location of industries, carrying out mining operations and regulating other activities in Doon Valley (February 1, 1989)
- Regulating activities in the coastal stretches of the country by classifying them as coastal regulation zones and prohibiting certain activities (February 19, 1991)
- Restricting location of industries and regulating other activities in Dahanu Taluka in Maharashtra (June 6, 91)
- Restricting certain activities in specified areas of Aravalli Range in the Gurgaon district of Haryana and Alwar district of Rajasthan (May 7, 1992)
- Restricting industrial and other activities, which could lead to pollution and congestion in the North West of Numaligarh in Assam (July 1996)

Purpose of Environment Impact Assessment

The different purposes of Environmental impact assessment are as follows:

Facilitation of Decision-Making: It provides a systematic evaluation of the environmental implications of a prepared action with or without alternatives prior to decision-making.

Aid in the process of development: Since the process of EIA provides a framework for the consideration of location, design and environmental issues besides the formulation of developmental actions for minimizing or eliminating the adverse impacts, many people see it as a hurdle in proceeding with different activities and also see the process involved in obtaining the permissions from various authorities as costly and time consuming (Annandale 2001).

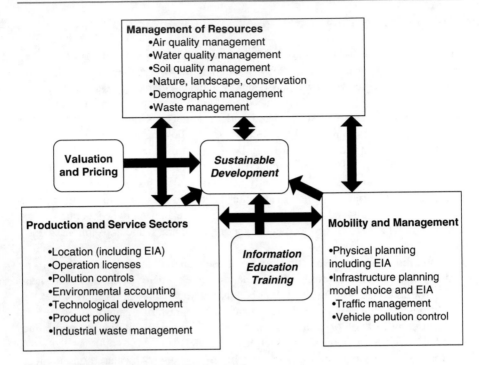

Fig. 3.1 Illustration of sustainable development

Acts as an instrument of sustainable development: As the maintenance of the overall quality of life, continued access to natural resources and avoiding of environmental damage are some key features of sustainable development, the institutional response to sustainable development at different levels are an important requirement. EIA as an institutional response mechanism works in these areas of any developmental activity well before the actual startup of the activity and hence aids in the process of sustainable development (Abaza et al. 2004).

As illustrated in Fig. 3.1, an interaction among the resources, sectors, and policies is necessary for sustainable development, and environmental impact assessment contributes specifically in that process.

Process of Environmental Impact Assessment

The Environmental Impact Assessment process is an iterative one, containing many feedback loops to allow the developmental proposal to be continually refined. So while the process of Environmental Impact Assessment follows a number of commonly accepted steps (Fig. 3.2), it does not observe a linear pattern.

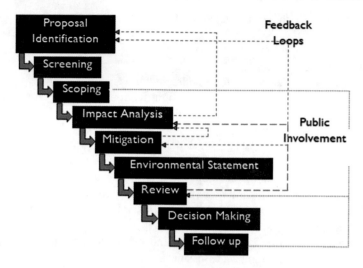

Fig. 3.2 Steps of environmental impact assessment process

Steps of Environmental Impact Assessment Process

Environmental impact assessment represents a systematic process that examines the environmental consequences of the developmental actions, in advance. The emphasis of environmental impact assessment is on prevention of inputs and therefore, is more proactive than reactive in nature. The environmental impact assessment process involves a number of steps listed below.

Proposal Identification
At the project identification and proposal development stage a number of decisions regarding project location, land use pattern, scale layout and design of the developmental project are made, the same can significantly reduce the impacts or in some cases can altogether reverse the impacts and can benefit the developer to minimize the mitigation liabilities at a later stage. Here we can take up the identification of the reasonable alternatives for the project and can evaluate the advantages and disadvantages of all the alternatives not only in terms of environment but also in terms of cost-effectiveness, feasibility, and reasonableness. The assessment of alternatives will result in the development of a preferred project proposal, which are then subjected to screening assessment.

Screening
Screening is undertaken to determine whether a developmental project requires an environmental impact assessment or not. The purpose of screening is to ensure that all developmental proposals likely to result in significant environmental impacts are subjected to an environmental impact assessment. Selection criteria that should be considered while screening a developmental proposal for environmental impact

assessment includes the characteristics of developmental location and potential effects.

Scoping
Scoping identifies the key issues to be addressed and ensures that the environmental impact assessment focuses on the areas of significant impacts, therefore preventing the use of resources in the areas of insignificant impacts. A good scoping process consists of three key components:

1. *Consultation*—with the relevant stakeholders and interest parties to provide them with the information on the developmental proposals and key issues and to find out their key concerns regarding the location of the same
2. *Analysis*—of the issues identified during *consultation* to determine their likely significance
3. *Negotiation*—with the decision makers and other interest parties to refine the scope of environmental impact assessment

Impact Analysis
It involves the characterization of impacts of a developmental proposal in terms of nature, spatiotemporal scale, duration, reversibility, frequency, and magnitude with a final judgment about their level of significance. Presently there exists a large number of impact analysis techniques and each one has its own advantages and disadvantages. Impact analysis techniques which can be qualitative or quantitative enable the analysis of likely impacts utilizing the existing as well as collecting the new information (Arts and Morrison 2004). Quantitative techniques tend to involve a strict method being set out and followed whereas the qualitative techniques rely less upon a prescribed method and instead rely heavily upon professional adjudgment.

Mitigation
The step that involves the development of those measures which help us in avoiding or reducing the likely significant environmental impacts of a developmental proposal is known as mitigation. Mitigation measures can be incorporated into the developmental design to get better results in terms of impact reduction. Further end of pipe mitigation measures can also be incorporated into the developmental process at a later stage.

Environmental Statement

It is a document used to communicate the results of the environmental impact assessment to the decision makers and to other stakeholders in the developmental process. It is a legal document that includes a nontechnical summary of the following information:

- Description of the project in terms of its site, size, and design
- Description of the impact reduction and remedial measures
- Description of the identified and assessed impacts on the environment
- Description of the main alternatives studied along with the reasons for their choice

The environmental statement, that is a clear and concise document should objectively enlist the environmental impact assessment process and findings, giving equal prominence to the positive and negative impacts relative to their importance. As the document is also used by many nonspecialists, any technical language should be avoided and if required the necessary technical information should be provided in a separate appendix (Arts and Morrison 2004).

Review

As the quality control is an important stage in any environmental impact assessment process, the review of the quality of environmental statement demonstrates whether the statement:

- Has met all the appropriate legal requirements
- Contains sufficient information to allow a decision to be made
- Is consistent with the current good practices

Review of the Environmental Statement can be undertaken at the draft stage or after finalization. The earlier the review takes place the greater the influence it can have over the quality of the Environmental Statement.

Decision-Making

A stage where the developmental proposal is either granted permission or not is known as decision-making. At this stage it is important to make the environmental statement available to a number of statutory committees, the general public, and other stakeholders. The decision makers also consider the current planning policies, socioeconomic information and the relevant local developmental frameworks and plans, as the final decision is based upon all of these things.

Follow-Up

It is an important stage which involves the actual implementation of the proposed mitigation measures proposed in the environmental statement. It is also the stage where any necessary monitoring of impacts is undertaken to ensure the proper implementation of the management plans which demonstrate a clear commitment to monitoring and mitigation measure (Arts and Morrison 2004).

Environmental Impact Assessment Methodology

The Environmental Impact Assessment methodology, that is a complex of procedures, tools and techniques helps in the fulfilment of the purpose of sustainable development (Ahmad and Sammy 1987) with the below mentioned methods as the commonly used ones. The purpose of Environmental Impact Assessment methodology is:

- To ensure the inclination of all the needed environmental factors in the process of analysis
- To provide means for evaluation of alternatives on a common basis
- To evaluate the mitigation measures, with a special focus to minimize the environmental impacts of alternatives and proposed action
- To provide information about public participation to ensure compliance to the govt. rules and regulations

Ad hoc Method
Ad hoc method involves a team of experts or specialists to identify impacts in their respective areas of expertise based on their trainings, intuitions, and combined experiences. It is not a real but a rough impact assessment method that does not actually structure a problem. It is a very easy method but has some major drawbacks:

1. All relevant impacts are not included.
2. Comparison of relative weightage of the different impacts is impossible.
3. It is inherently inefficient as it requires sizeable efforts to assemble and identify an appropriate panel of experts for each assessment.
4. It provides minimal guidance for impact analysis while suggesting broader areas of possible impacts.

Overlays
Overlays is based on the use of a series of transparent (overlay) maps with each map representing the landscape and other environmental factors (Fig. 3.3). These maps are overlaid to produce a composite map to characterize the physical, biological, ecological, social, and other relevant features of the area relative to the location of the proposed development. Information is collected for the standard geographical units and recorded on a series of such maps, typically one for each variable. The validity of the analysis is related to the number and type of parameters chosen to investigate the degree of associated impacts. Any number of project alternatives can be located on the final map; however, for a readable composite map, the number of parameters in the transparency is usually restricted to ten (Morris and Therivel 2001).

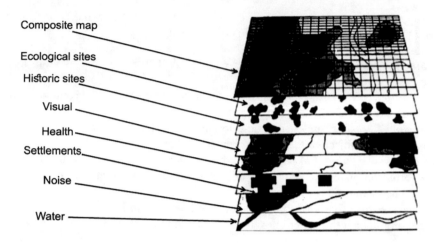

Composite map

Ecological sites

Historic sites

Visual

Health

Settlements

Noise

Water

Fig. 3.3 Overlay maps

The method that is spatially oriented and is capable of communicating the spatial aspects of the cumulative impacts in a clear fashion has a number of benefits and limitations.

Benefits

1. It is conceptually simple and provides a clear presentation.
2. It is highly versatile and appropriate for assessing impacts occurring in large areas.
3. It helps in predicting the geographical location of impacts.
4. It is very useful in the development of alternative sites.

Limitations

1. It lacks the casual explanation of the impact pathways.
2. It lacks the production capability with respect to the population effects; however, some sophisticated versions can make some predictions about habitat loss.

Checklists

A standard list of different impacts associated with a specific action is known as a checklist. These methods organize information and ensure that no potential impact is overlooked. Checklists are basically modified versions of the ad hoc method that involve the listing of the specific areas of impacts and supplies the instruction for impact identification and evaluation (Pendse et al. 1989). Checklist is one of the basic methodologies used in EIA and is of following types:

1. *Simple Checklist:* list of parameters with no guidelines on how they are to be measured and interpreted
2. *Descriptive Checklist:* includes an identification of environmental parameters and guidelines on how to measure the data on a particular parameter
3. *Scaling Checklist:* similar to a descriptive checklist with additional information basic to the subjective scaling of parameter values
4. *Scaling Weighting Checklist:* similar to a scaling checklist, with additional information about the subjective evaluation of each parameter with respect to all other parameters
5. *Questionnaire Checklist:* based on a set of questions to be answered with some questions about the direct impacts and possible mitigation measures

Checklists are used for the following main reasons:

1. They are useful in summarizing information to make it accessible to specialists from other fields, or to decision makers who may have a limited technical knowledge about the same.
2. They provide a preliminary level of analysis and a mechanism for incorporating information about ecosystem functions.

Westman (1985) listed the following few problems using checklists as an impact assessment method:

1. Their generalistic and incomplete nature
2. Their lack of illustrative intentions between effects
3. Their distraction from the most significant impacts
4. Qualitative and subjective identification of effects

Benefits of Checklists

1. Checklists are comparatively simple.
2. Not necessarily project specific.
3. Once established, can be used in many different situations.

Networks

Given by Sorenson in 1971, Network method primarily explains the linkages between different environmental aspects and is solely used to illustrate the primary, secondary, and tertiary impacts of a developmental activity on different components of environment. These methods establish a cause–condition–effect relationship between the developmental actions, the environmental parameter, and the impacts.

Networks establish the causal chain of impacts offering three dimensional approaches to identify impacts and can be limited by minimal information on technical aspects (Clark et al. 1978). These methods are usually in the form of flowcharts or diagrams, as illustrated in Fig. 3.4.

After the identification of all the primary impacts of a developmental activity, the primary impacts are used to identify the secondary and tertiary impacts and thus connected into the network. In developing a network diagram, the first step is the identification of first-order impacts followed by identification of second- and third-order impacts and this process continues until the network diagram is completed to the practitioner's satisfaction. The networks help in exploring and understanding the underlying relationships between environmental components that produce higher order changes, often overlooked by simpler approaches.

Advantages

1. Help to follow the chain of events and the associated impacts of a developmental action
2. Assessment of multiple order impacts at a time
3. Identification of links overlooked in other methods
4. Easy and esthetically pleasing

Disadvantages

1. Unlike matrices networks give no information about the significance and magnitude of impacts
2. Networks, at times get long and messy especially in case of large-scale projects
3. Deeper knowledge of the environment conditions of the project area is required

Leopold Matrix

Leopold et al. (1971) designed a matrix with some hundred specified actions and 88 environmental components (Table 3.3), considering each action for its potential impacts on each environmental parameter. The magnitude of the interaction (extensiveness or scale) is described by assigning a value ranging from 1 (for small magnitudes) to 10 (for large magnitudes). The assignment of numerical values is based on an evaluation of available facts and data. Similarly, the scale of importance also ranges from 1 (very low interaction) to 10 (very important interaction). Assignment of numerical values for importance is based on the subjective judgment of the interdisciplinary team working on the environmental impact assessment study.

One of the most attractive features of the Leopold Matrix is that it is reasonably flexible and the total number of specified actions and environmental items may increase or decrease depending on the nature and scope of the study and the specific "Terms of References" for which the environmental impact study is undertaken (Leopold et al. 1971).

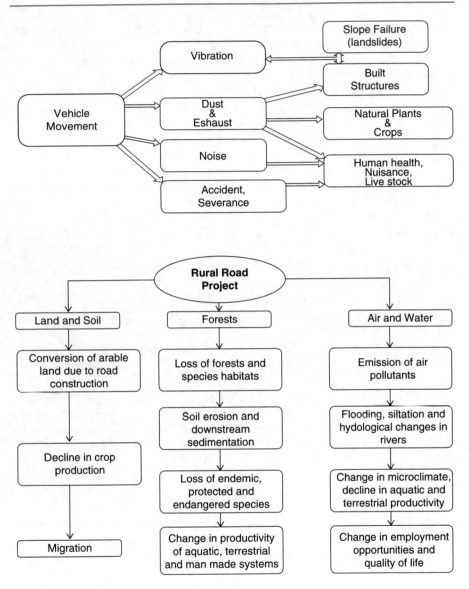

Fig. 3.4 Use of networks method in a road sector example

Technically, the Leopold Matrix approach is a gross screening technique to identify impacts. It is a valuable tool for explaining impacts by presenting a visual display of the impacted items and their causes. Summing the rows and columns that are designated as having interactions can provide a deeper insight and aid in further interpretation of the impacts. The matrix can also be employed to identify impacts

Table 3.3 Actions and environmental items in the Leopold Matrix

	Actions		Environmental items
Category	Description	Category	Description
		A. Physical & chemical characteristics	
A. Modification of regime	Exotic fauna introduction (a) Biological controls (b) Modification of habitat (c) Alteration of ground cover (d) Alteration of groundwater hydrology (e) Alteration of drainage (f) River control & flow modification (g) Canalization (h) Irrigation (i) Weather modification (j) Burning (k) Surface or paving (m) Noise & vibration	1. Earth	(a) Mineral resources (b) Construction material (c) Soils (d) Land form (e) Force fields & background radiation (f) Unique physical features
B. Land transformation & construction	(a) Urbanization (b) Industrial sites & buildings (c) Airports (d) Highways & bridges (e) Roads & trails (f) Railroads (g) Cables & lifts (h) Transmission lines, pipelines & corridors (i) Barriers including fencing (j) Channel dredging & straightening (k) Channel retaining walls (l) Canals (m) Dams & impoundments (n) Piers, seawalls, marinas & sea terminals (o) Offshore structures (p) Recreational structures (q) Blasting & drilling (r) Cut & fill (s) Tunnels & underground structures	2. Water	(a) Surface (b) Ocean (c) Underground (d) Quality (e) Temperature (f) Recharge (g) Snow, ice, & permafrost

(continued)

Table 3.3 (continued)

	Actions		Environmental items	
Category	Description	Category	Description	
C. Resource extraction	(a) Blasting and drilling (b) Surface excavation (c) Subsurface excavation & retorting (d) Well dredging & fluid (e) Dredging (f) Clear cutting & other lumbering (g) Commercial fishing & hunting	3. Atmosphere	(a) Quality (gases, particulates) (b) Climate (micro, macro) (c) Temperature	
D. Processing	(a) Farming (b) Ranching & grazing (c) Feed lots (d) Dairying (e) Energy generation (f) Mineral processing (g) Metallurgical industry (h) Chemical industry (i) Textile industry (j) Automobile & aircraft (k) Oil refining (l) Food (m) Lumbering (n) Pulp & paper (o) Production shortage	4. Processes	(a) Floods (b) Erosions (c) Deposition (sedimentation, precipitation) (d) Solution (e) Sorption (ion exchange, complexing) (f) Compaction & settling (g) Stability (slides, slumps) (h) Stress-strain (earthquakes) (i) Air movements	
		B. Biological conditions		
E. Land alteration	(a) Mine sealing and waste control (b) Strip mining rehabilitation (c) Landscaping (d) Harbor dredging (e) Marsh fill and drainage	1. Flora	(a) Shrubs (b) Grass (c) Crops (d) Micro flora (e) Aquatic plants (f) Endangered species (g) Barriers (h) Corridors	
F. Resource renewal	(a) Reforestation (b) Wildlife stocking and management (c) Groundwater recharge (d) Fertilization application (e) Waste recycling	2. Fauna	(a) Birds (b) Land animals including reptiles (c) Fish & shellfish (d) Benthic organisms (e) Insects (f) Microfauna (g) Endangered species (h) Barriers	
		C. Cultural factors		

(continued)

Table 3.3 (continued)

Actions		Environmental items	
Category	Description	Category	Description
G. Changes in traffic	(a) Railway (b) Automobile (c) Trucking (d) Shipping (e) Aircraft (f) River and canal traffic (g) Pleasure boating (h) Trails (i) Cables and lifts (j) Communication (k) Pipeline	1. Land use	(a) Wilderness and open spaces (b) Wetlands (c) Forestry (d) Grazing (e) Agriculture (f) Residential (g) Commercial (h) Industry (i) Mining and quarrying
H. Waste replacement & treatment	(a) Ocean dumping (b) Landfill (c) Emplacement of tailings, spoils and overburden (d) Underground storage (e) Junk disposal (f) Oil well flooding (g) Deep well emplacement (h) Cooling water discharge (i) Municipal waste discharge (j) Liquid effluent discharge (k) Stabilization and oxidation ponds (l) Septic tanks, commercial and domestic (m) Stack and exhaust emission (n) Spent lubricants	2. Recreation	(a) Fishing (b) Boating (c) Swimming (d) Camping and hiking (e) Picnicking (f) Resorts
		3. Esthetic & human interest	(a) Scenic views and vistas (b) Wilderness qualities (c) Open-space qualities (d) Landscape design (e) Unique physical features (f) Parks and reserves (g) Monuments (h) Rare and unique species or eco-systems (i) Historical or archaeological sites and objects (j) Presence of misfits
I. Chemical treatment	(a) Fertilization (b) Chemical deicing of highways, etc. (c) Chemical stabilization of soil (d) Weed control (e) Insect control (pesticides)	4. Cultural status	(a) Cultural patterns (lifestyle) (b) Health and safety (c) Employment (d) Population density
J. Accidents	(a) Explosions (b) Spills and leaks (c) Operational failure	5. Manufactured facilities and activities	(a) Structures (b) Transportation network (movement, access) (c) Utility networks

(continued)

Table 3.3 (continued)

Category	Actions Description	Category	Environmental items Description
			(d) Waste disposal
			(e) Barriers
			(f) Corridors
K. Others		D. Ecological relationships	(a) Stalinization of water resources
			(b) Eutrophication
			(c) Disease-insect vectors
			(d) Food chains
			(e) Stalinization of surficial material
			(f) Brush encroachment
			(g) Other
		E. Others	

Source: Canter (1977)

during the various parts of the entire project cycle—construction, operation, and even dismantling phases.

Quantitative or Indexed Methods

Environmental Impact Assessment Guidelines 2006 and Amendments

In April 2006 the Ministry of Environmental Affairs and Tourism passed environmental impact assessment regulations (the Guidelines) in terms of the National Environmental Management Act, 1998 (NEMA). The Guidelines replaced the environmental impact assessment (EIA) guidelines (1997), which were passed earlier in terms of the Environment Conservation Act, 1989. In order to assist potential applicants, environmental assessment practitioners ("EAPs") and interested and affected parties ("I&APs") to understand what is required of them in terms of the Regulations, what their rights are and/or what their role may be, the Department of Environmental Affairs and Tourism has expanded its Integrated Environmental Management Guideline Series to include the following documents:

- Guideline 3: General guide to the EIA Regulations
- Guideline 4: Public participation
- Guideline 5: Assessment of alternatives and impacts
- Guideline 6: Environmental management frameworks

The Ministry of Environment Forests (MoEF), GoI issued the Environmental Impact Assessment Notification on September 14, 2006, which makes prior environmental clearance mandatory for a number of developmental activities listed in its schedule.

Prior Environmental Clearance Procedure

Step 1: Application for Prior Environmental Clearance

The following projects or activities require prior environmental clearance from the concerned regulatory authority, before starting any construction work, or preparation of land by the project management (except fencing the land):

1. All new projects or activities listed in the schedule of EIA notification, 2006 such as mining of minerals, river valley projects, thermal power projects, and cement industries etc.
2. Expansion and modernization of existing projects or activities listed in the schedule with addition of capacity beyond the threshold limits specified for the concerned sector
3. Any change in product—mix in an existing manufacturing unit included in the schedule beyond the specified range

All projects and activities are broadly categorized into two categories—Category A and Category B, based on the spatial extent of potential impacts on human health, natural and man-made resources. As per the categorization of projects (A or B) done in the schedule of EIA Notification, applicant should submit the proposal to Central Government or State Authority.

The Category "A" projects/activities require prior environmental clearance from the Ministry of Environment and Forests (MoEF) on the recommendations of Expert Appraisal Committee (EAC) constituted by the Central Government. At State level the State Environment Impact Assessment Authority (SEIAA) is the regulatory authority for matters falling under Category "B," which receives projects recommended by State level Expert Appraisal Committee (SEAC). The SEAC and SEIAA are constituted by Central Government.

Step 2: Project Appraisal at SEAC

Received applications are processed by SEAC in four stages viz. screening, scoping, public consultation, and appraisal. Next to appraisal, the SEAC recommends the project to SEIAA, along with its suggestions about issuing "Environmental Clearance" or rejecting the application.

Stage (1): Screening

- This stage involves the scrutiny of application by SEAC for Category "B" projects or activities. SEAC determines whether or not the project or activity requires further environmental studies for preparation of an Environmental Impact Assessment (EIA), depending upon the nature and location specificity of the project. Projects requiring an EIA report are termed as Category "B1" while the remaining ones are termed as Category "B2."

Stage (2): Scoping

- Scoping refers to the process by which the EAC (for Category "A" projects or activities), and SEAC (for Category "B1" projects or activities) determines the detailed and comprehensive Terms of Reference (TOR) addressing all relevant environmental concerns for the preparation of an Environment Impact Assessment (EIA) Report in respect of the project or activity for which prior environmental clearance is sought.
- Applications for prior environmental clearance may be rejected by the regulatory authority concerned (MoEF or SEIAA) on the recommendations of the EAC or SEAC concerned even at this stage and the decision together with reasons for the same has to be communicated to the applicant in writing within 60 days of the receipt of application.

Stage (3): Public Consultation

- Public Consultation refers to the process by which the concerns of local affected persons and others who have plausible stake in the environmental impacts of the project or activity are ascertained with a view to account all the material concerns in the project or activity design as appropriate.
- All Category "A" and Category "B1" projects or activities are required to undertake Public Consultation, except the following:
  1. Modernization of irrigation projects
  2. All projects or activities located within industrial estates or parks approved by the concerned authorities, and which are not disallowed in such approvals
  3. Expansion of Roads and Highways which do not involve any further acquisition of land
  4. All Building/Construction/Area Development projects and Townships
  5. All Category "B2" projects and activities
  6. All projects or activities concerning national defense and security or involving other strategic considerations as determined by the Central Government
  The Public Consultation ordinarily comprises two components:

1. A public hearing at the site or in its close proximity (district wise) to be carried out in the manner prescribed in the notification, for ascertaining concerns of local affected persons
2. Obtain responses in writing from other concerned persons having a plausible stake in the environmental aspects of the project or activity:
   - The public hearing at, or in close proximity to, the site(s) in all cases is conducted by the State Pollution Control Board (SPCB) or the Union territory Pollution Control Committee (UTPCC) concerned in the specified manner and forward the proceedings to the regulatory authority concerned within 45 days of a request to the applicant.
   - After completion of the public consultation, the applicant shall address all the material environmental concerns expressed during this process, and make appropriate changes in the draft EIA and EMP. The final EIA report, so prepared, has to be submitted by the applicant to the concerned regulatory authority for appraisal or the applicant may alternatively submit a supplementary report to draft EIA and EMP addressing all the concerns expressed during the public consultation.

Stage (4): Appraisal

- Appraisal means the detailed scrutiny by the EAC or SEAC of the application and other documents like the Final EIA report, outcome of the public consultations including public hearing proceedings, submitted by the applicant to the regulatory authority concerned for grant of environmental clearance.
- The appraisal has to be made by EAC or SEAC concerned in a transparent manner in a proceeding to which the applicant is invited for furnishing necessary clarifications in person or through an authorized representative.
- On conclusion of the proceedings, the EAC or SEAC concerned makes categorical recommendations to the regulatory authority concerned either for grant of prior environmental clearance on stipulated terms and conditions, or rejection of the application for prior environmental clearance, together with reasons for the same.
- The appraisal of an application has to be completed by the EAC or SEAC concerned within 60 days of the receipt of the final Environment Impact Assessment report and other documents or the receipt of Form 1 and Form 1A, where public consultation is not necessary and the recommendations of the EAC or SEAC has to be placed before the competent authority for a final decision within the next 15 days.

The members of the EAC and SEAC, concerned, may inspect any site(s) connected with the project or activity in respect of which the prior environmental clearance is sought, for the purposes of screening or scoping or appraisal.

Step 3: Environmental Clearance by SEIAA

A State Level Environment Impact Assessment Authority (SEIAA) is constituted by the Central Government under the Environment (Protection) Act, 1986 as the Environmental Clearance issuing authority, which scrutinizes the projects appraised by SEAC and gives the final decision on issuing Environmental Clearance or rejecting the application.

Environment Impact Statement (EIS)

Environmental Impact Statement (EIS) is a document prepared to describe the effects of proposed activities on the environment. An EIS describes impacts, as well as ways to mitigate them. It also describes impacts of alternatives as well as plans to mitigate the impacts.

A typical EIS contains the following three parts:

Part 1—Methods and key issues: This part deals with the statement of methods used and a summary of key issues.

Part 2—Background to the proposed development: This part deals with preliminary studies (i.e., need, planning, alternatives, site selection, etc.), site description/ baseline conditions and description of proposed development, construction activities, and programs.

Part 3—Environmental impact assessments on target areas: This part deals with land use, landscape and visual quality, geology, topography and soils, hydrology and water quality, air quality and climate, terrestrial and aquatic ecology, noise, transport, socioeconomic and interrelationships between effects.

EIS Procedure

The EIS is prepared in a series of steps involving *gathering government and public comments to define the issues that should be analyzed in the EIS (a process known as "scoping")*; preparing the draft EIS; *receiving and responding to public comments on the draft EIS*; and preparing the final EIS (Fig. 3.5). Decisions are not made in the EIS; rather, the EIS analysis is one of the several factors that the decision makers consider and the decision is announced in the Record of Decision (ROD) after the final EIS has been published (Eccleston 2014).

Fig. 3.5 Steps in EIS process

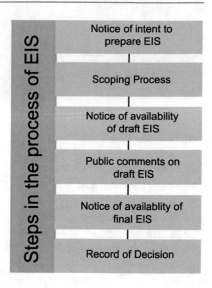

Scoping

A federal agency begins the scoping process for an EIS by publishing a Notice of Intent (NOI) in the Federal Register to let the public know that it is considering an action and in preparing an EIS. The NOI describes the proposed action and provides background information on issues and potential impacts. During the scoping period, the public can provide comments on the proposed action, alternatives, issues, and environmental impacts to be analyzed in the EIS. Scoping may involve public meetings and other means to obtain public comments on the EIS.

Draft EIS

Preparation of a draft EIS presents, analyzes, and compares the potential environmental impacts for the proposed action and alternatives and their implementation is the next step in the process. It provides additional information on the methodologies and assumptions used for the analyses. A Notice of Availability (NOA) is published in the Federal Register *announcing the availability of the draft EIS for public review and comment*. The NOA takes a minimum 45-day comment period and the public comments on the draft EIS are considered in the preparation of the final EIS.

Final EIS

Once the public comment period on the draft EIS has been completed, a final EIS is prepared by incorporating the public comments on the draft EIS are included in the final EIS into it and is subsequently distributed.

Record of Decision

After the publication of final EIS, a minimum 30 day waiting period is required before a record of decision is issued. The record of decision notifies the public about the decision made on the proposed action and presents the reasons for that decision. The decision-making process includes consideration of factors such as cost, technical feasibility, agency statutory missions, and national objectives, as well as the potential environmental impacts of the action(s). No action can be taken until the decision has been made public (Eccleston 2008).

Strategic Environmental Assessment (SEA)

One of the most recent trends in the application of EIA at earlier and more strategic stages of development like the level of policies, plans, and programs, is known as strategic environmental assessment (SEA). It is also defined as the formalized, systematic, and comprehensive process of evaluating the environmental impacts of a policy, plan, or program (PPP) and its alternatives, including the preparation of a written report on the findings of that evaluation, and using the findings in publicly accountable decision. In other words, the EIA of policies, plans, and programs, keeping in mind that the process of evaluating environmental impacts at a strategic level, is not necessarily the same as that at a project level. In theory, policy, plan, or programs are tiered—a policy provides a framework for the establishment of plans, plans provide frameworks for programs, and programs lead to projects (Abaza et al. 2004). The EIAs for these different policy, plan, or program tiers can themselves be tiered as shown in Fig. 3.6 and so the issues at higher tiers need not be reconsidered as the lower tiers:

A hierarchy exists between policies, plans, and programs with policies at the top level of conceptualization (Fig. 3.7) and generality; plans at one level down the policies, and above the programs. Programs make plans more specific by including a time schedule for specific activities. Implementation of a program involves carrying out specific projects, which can be subjected to traditional EIA (Lyer 2017).

Rationale and Scope of SEA

Strategic Environmental Assessment (SEA) is a set of analytical and participatory processes for incorporating environmental considerations, at early stages of

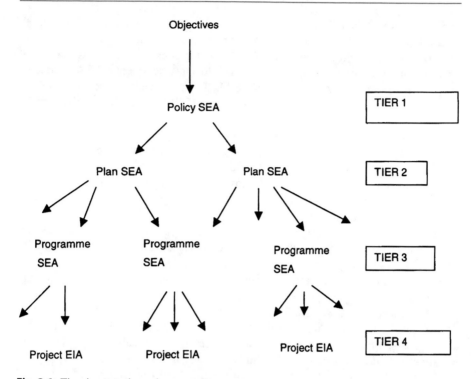

Fig. 3.6 Tiers in strategic environmental assessment

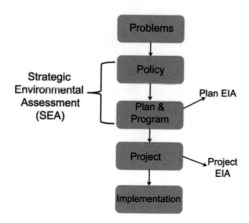

Fig. 3.7 Policy Plan Program (PPP) and SEA

decision-making, into policies, plans, and programs that affect natural resources. SEA evaluates, at the highest strategic level, a project's environmental impacts in the context of social and economic factors. This extends traditional Environmental Impact Assessments (EIA), which focuses solely on physical environmental impacts. SEA fosters and provides critical systematic considerations at the sectoral, regional, and national levels to promote environmental sustainability, smart growth, and pollution prevention. The term "SEA" encompasses a spectrum of analytical processes such as Strategic Environmental and Social Assessment (SESA), Strategic Social Environmental Assessment (SSEA), Country Environmental Analysis (CEA), Combined Strategic Impact Assessment (CSIA), and Cumulative Impact Assessment (CIA). Strategic Environmental Assessment (SEA) is a process that ensures that significant environmental impacts arising from policies, plans, and programs are identified, assessed, mitigated, communicated to decision makers, and that opportunities for public involvement are provided (Lyer 2017). SEA has a key role to play in delivering the purpose of sustainable economic growth, by ensuring greater consideration of the impacts and by giving a greater opportunity to public participation in decision-making.

In broader terms, the rationale for SEA of policies, plans, and programs falls into three main categories: strengthening project EIA, advancing the sustainability agenda, and addressing cumulative and large-scale effects.

SEA, or an equivalent approach, can be used as a complement to project-level EIA to incorporate environmental considerations and alternatives directly into policy, plan, and program design. Thus, when applied systematically in the upstream part of the decision cycle and to the economic, fiscal, and trade policies that guide the overall course of development, SEA can be a vector for a sustainability approach to planning and decision-making (Brundtland Commission of WCED, 1987 and UNCED, 1992). This upstream approach can also help in making EIA projects more consequential and reducing the time and effort involved in their preparation. SEA may yield other significant benefits as well, e.g., by ruling out certain kinds of development at the policy level, reducing the need for many project-level EIA and thus relieving pressure where institutional and/or skill capacity is limited.

Arguably, SEA offers a better opportunity than project-level impact assessment to address cumulative impacts. Recently, considerable efforts have been made to extend EIA-based frameworks to encompass certain types of cumulative impacts. These deal reasonably well with the ancillary impacts of large-scale projects (e.g., dams and transport infrastructure) and the incremental effects of numerous, small-scale actions of a similar type (e.g., road realignment and improvement). However, more pervasive cumulative effects and large-scale environmental changes are difficult to address. In principle, these can be addressed best by SEA of policies, plans and program; however, this has not proven to be the case in practice.

Scope of SEA

Most practitioners view SEA as a decision aiding rather than a decision-making tool. In other words, it is seen as a tool for forward planning to be flexibly applied at various stages of the policy-making cycle. Under this broad perspective, SEA encompasses assessment of both broad policy initiatives and more concrete programs and plans that have physical and spatial references. With the scope of coverage, one problem that becomes evident is the methodologies to be applied at the opposite ends of the decision-making spectrum differ markedly. However, the principles of EIA apply at all levels.

Comparison of EIA and SEA

EIA	SEA
Is usually reactive to a developmental proposal.	Is proactive and informs developmental proposals.
Assesses the effect of a proposed development on the environment.	Assesses the effect of a policy, plan, or program on the environment, or the effect of the environment on development needs and opportunities.
Addresses a specific project.	Addresses areas, regions, or sectors of development.
Assesses direct impacts and benefits.	Assesses cumulative impacts and identifies implications and issues for sustainable development.
Focuses on the mitigation of impacts.	Focuses on maintaining a chosen level of environmental quality.
Narrow perspective and a high level of detail.	Wide perspective and a low level of detail to provide a vision and overall framework.
Focuses on project-specific impacts.	Creates a framework against which impacts and benefits can be measured.

SEA Process

In project EIA, impact mitigation, i.e., avoiding or reducing the impacts of the project, restoring the affected environment or compensating for adverse effects, is often considered as a separate stage in the process. In SEA, instead, the focus of the project is on reconsidering the policy, plan, or program (PPP) from a cross-cutting perspective, leading to an improved understanding of the policy, plan, or program and possibly changes to the policy, plan, or program: each stage considers whether and how the policy, plan, or program can be changed and improved. These changes mostly involve rewriting the policy, plan, or program to minimize any negative environmental/sustainability impact. They also involve establishing management guidelines for the implementation of the policy, plan, or program, placing constraints on lower-tier policy, plan, or program or developing environmentally beneficial shadow policy, plan, or programs or projects (Lyer 2017). As such, mitigation in SEA is an ongoing process as illustrated in Fig. 3.8.

An SEA process involves the following stages:

1. *Screening:* At this stage, responsible agencies carry out an appropriate assessment of all strategic decisions with significant environmental consequences.

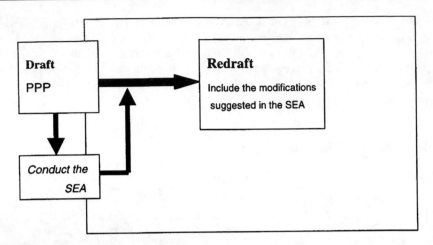

Fig. 3.8 Mitigation of SEA

2. *Timing:* At this stage, results of the assessment are available sufficiently early for use in the preparation of the strategic decision.
3. *Environmental scoping:* At this stage, all relevant information is provided to judge whether an initiative should proceed and objectives could be achieved in a more environmentally friendly way (i.e., through alternative initiatives or approaches).
4. *Other factors:* At this stage, sufficient information is available on other factors, including socioeconomic considerations, either parallel to, or integrated in, the assessment.
5. *Review:* At this stage, the quality of the process and information is safeguarded by an effective review mechanism.
6. *Participation:* At this stage, sufficient information on the views of all legitimate stakeholders (including the public affected) is available early enough to be used effectively in the preparation of the strategic decision.
7. *Documentation:* At this stage, results are identifiable, understandable, and available to all parties affected by the decision.
8. *Decision-making and accountability:* At this stage, it should be clear to all stakeholders and all parties affected how the results are taken into account in decision-making.
9. *Post-decision:* At this stage, sufficient information on the actual impacts of implementing the decision is gained to judge whether or not the decision should be amended.

Methodological Difference Between EIA and SEA

• Scale of SEA is wider than EIA as it involves number of activities, assessment of a larger extent of impacts, deals with a greater range of alternatives and a wider area of significance.

- Time interval is longer in SEA between planning, approval, and implementation. Even data collection in SEA is at a time-consuming stage.

Benefits and Constraints of SEA

Benefits

1. Promoting integrated environmental and developmental decision-making
2. Facilitating the design of environmentally sustainable policies and plans
3. Providing for consideration of a wide range of alternatives than is normally possible in project EIA
4. Accounting of cumulative effects and global change
5. Enhancing institutional efficiency (particularly, where EIA-related skills, operational funds and institutional capacities are limited) by obviating the need for unnecessary project-level EAs
6. Strengthening and streamlining project EA by incorporating environmental goals and principles into policies, plans, and programs that shape individual projects, identifying in advance the impacts and information requirements, resolving strategic issues and information requirements and reducing time and effort taken to conduct reviews
7. Providing a mechanism for public engagement in discussions relevant to sustainability at a strategic level

Constraints

- A level of institutional maturity is necessary, which allows for effective inter-sectoral dialog, for environmental considerations to be taken into account in formulating, revising, and implementing policies, plans, and programs effectively, and to influence decision-making.
- Appropriate skills are needed, within government departments/agencies and private sectors (e.g., industry, environmental consulting companies) and amongst academics and NGOs.

In practice, the extents to which the benefits of SEA are achieved also depend on a number of other important factors such as:

- Provisions made for SEA, e.g., legal versus administrative
- Prior record of implementation and acceptance by decision makers
- Degree to which overall strategies of sustainable development are in place
- Scope and level(s) of process application with the broadest range of benefits being gained from SEA systems that include review of policies as well as plans and programs (Donnelly et al. 2006).

Public Participation in Environmental Decision-Making

Public means one or more natural or legal persons and, in accordance with national legislation or practice, their associations, organizations or groups.

Although, the meaning of public participation is difficult to articulate, it has been defined as *"purposeful activities in which citizens take part in relation to government."* It has also been described to be composed of four elements:

- The purposes for which the participation is undertaken
- The type of action that is undertaken
- The individuals who are involved in the action and
- The governmental entities that are targeted

Environmental Decision-Making

- Any process of decision-making where significant environmental impacts are a possibility.
- Includes law making, policy-making, spatial planning, strategic planning, resource management planning, licensing of industry, etc.
- Studies indicate a serious gap in understanding and trust between stakeholders.
- Existing decision-making structures are often based on either the traditional DAD (Decide, Announce, and Defend) or DEAD (Decide, Educate, Announce, and Defend) methods, and this can be seen as one source of this gap in trust.

Objective of Public Participation

Public participation in decision-making is a core issue of good environmental governance. Participation rights and representation, as well as accountability and transparency are among seven key elements of environmental governance. Public participation at different levels raises accountability and reliability of decisions, lessens risks of possible conflicts and inconsistencies and facilitates implementation (Lee and Abbot 2003).

Public participation in decision-making is an essential part of the environmental impact assessment (EIA) process, which has become a widely applicable tool for environmental decision-making in the world since 1970s, ensuring consideration of environmental concerns within the planning. Different countries practice different levels of public involvement. Although, more successful democracies have got much forward in this sense, some of the newly emerged democracies have introduced EIA systems just recently.

Public Participation in Decision-Making

According to the Aarhus convention (UNECE 1998), following are a few domains for public participation in decision-making:

- Participation in decisions on specific activities
- Participation concerning plans, programs, and policies relating to the environment
- Participation during the preparation of executive regulations and/or generally applicable legally binding normative instruments

Mechanism for Public Participation in Decision-Making

Public Hearing

Proactive role of public affirms that responsible civil society participation in partnership with government is a reflection of good governance and environmental issues are best handled with the participation of all concerned citizens at the relevant levels. Full and effective participation of women, tribals and other traditionally marginalized groups is essential to achieve sustainable development. The strength and meaning of representative democracy lie in the active participation of individuals at all levels of civil life.

According to Ministry of Environment and Forests (MOEF), for Effective Public Participation under Provision of Environmental Protection Act, as a part of EIA notification dated April 10, 1997, a public hearing is mandatory for 29 categories of Projects which belongs to EIA amendment notification of 2006 and as per the same notification mechanism of public participation involves:

- Access to Information—information on relevance of policy formulations, performance of regulated entities, EIAs and other important information
- Access to Process—process of making policies, laws and regulations and those of granting permits, limits and other conditions
- Access to Justice—access to courts, administrative appeals and other relevant decision-making bodies/tribunals

Salient features of the notification for public hearing are:

- 30 days prior notice to public regarding decision-making for environmental issues
- Notice in Press to inform the public
- Notice to mention date, time, and venue of the meeting
- Persons can give oral/written objections
- Executive Summary to be made available to the public
- Composition of Hearing Panel fixed—with amended notification

Composition of Panel for Public Hearing

Following person/officials are the compositions of public hearing:

- *District Collector*—Chairman
- Representative of state pollution control board
- Representative of Government dealing with the subject
- Representative of the State Government dealing with Environment
- Not more than three representatives of local bodies such as municipalities/ panchayats
- Not more than three senior citizens of the area nominated by the District Collector.

Environmental Auditing (EA)

An environmental audit (EA) is a systematic, independent internal review to check whether the results of environmental work tally with the targets. It studies whether the methods or means used to achieve the goals or ends are effective. EA involves studying documents and reports, interviewing key people in the organization, etc. to assess the level of deviations between targets and results.

It is defined as a systematic and documented verification process of objectively obtaining and evaluating evidence to determine whether an organization's EMS conforms with audit criteria set by the organization and for communicating the results of this process to management (ISO 14001).

An environmental audit is being used as a tool and as an aid to test the effectiveness of environmental efforts at local level. It is carried out for a number of reasons including the following:

- To verify compliance
- To review implementation of policies
- To identify liabilities
- To review management systems
- To identify needs, strengths, and weaknesses
- To assess environmental performance
- To promote environmental awareness

Objectives

The objectives of an environmental audit are to evaluate the efficiency and efficacy of resource utilization (i.e., people, machines, and materials), to identify the areas of risk, environmental liabilities, weakness in management systems and problems in complying with regulatory requirements and to ensure the control on waste/pollutant generation (Barton and Bruder 1995).

Environmental audit deals in the following areas:

Design Specification and Layout While setting up an industry, adequate provisions are made in the design specification and layout to augment the production capacity but corresponding provisions to meet the environmental criteria are often overlooked. Adequate provisions are, therefore, necessary to upgrade pollution control measures to meet the future environmental standards that are getting stringent day by day (Young 1994). The audit helps in identifying specific areas of concern to meet the future requirements of environmental measures.

Resource Management The audit provides data for the efficient use of the resources include air, water, and energy per unit production, and, thereby, helps to reduce resource consumption and waste production.

Pollution Control Systems and Procedures The audit helps us to ensure that the systems and procedures governing the environmental activities/operations of pollution control equipment are rightly followed and determines the efficiency of the system in identifying conditions and inviting corrective actions in a timely and effective manner.

Emergency Plans and Response/Safety System As the emergency plans more often than not remain in the safe custody of senior management, staff may not have immediate access to the right action during an emergency. The review of the emergency response system ensures adequate knowledge, alertness, and readiness of the staff concerned to effectively face an emergency.

Medical and Health Facilities/Industrial Hygiene and Occupational Health The productive element of an industry is dependent on the health of its human resources. The primary facilities to suit the occupational needs of the industry are, therefore, vital. Audit in this regard provides an insight into the actual requirements to warn suitable orientation of existing facilities.

Confirmation to Regulatory Requirement The regulatory mechanism of environmental compliance is gradually becoming more and more comprehensive. New regulations and standards are being stipulated at such a pace that they render the existing systems archaic. Factory managers not being fully aware of the latest requirements make the top management/owners vulnerable for prosecution under various environmental acts. An audit helps to compare the existing status with the stipulation and standards prescribed by various agencies and ensures their compliance.

Scope

The scope of environmental audit is very flexible, depending on the needs of the organization. An environmental audit should be conducted in a manner which allows for provision of information regarding:

- The history of an organization or activity, including information on the setting, previous environmental damage at the site, environmental practices, monitoring records and known environmental issues
- The natural resources used as input, processing of materials and all finished products and wastes including hazardous and toxic wastes
- The handling and storage of chemicals, hazardous and toxic materials and any potential environmental hazards
- Environmental risk assessment
- Waste management control systems, transportation route for materials and waste disposal, including facilities to minimize waste disposal impacts and accidents
- Measure of the effectiveness of pollution control equipment as indicated in inspection reports, maintenance logs, emission test results, and routine analytical reports
- Emergency response plans and procedures
- Recycling, programs, and product life cycle considerations
- Plans to increase environmental awareness

Steps Involved in Environmental Audit

Four general steps involved in an audit procedure are:

1. *Audit preparation:* This includes choice of auditor/audit group, collection of background material and planning of the audit orientation. Audit preparation is crucial in determining the methodology, practical tools and materials that are required for the completion of the audit. In many cases, the available materials such as manuals, questionnaires, etc., may not be well suited to the type of activity that is to be audited, or the type of audit that is to be performed. This means that part of the preparatory activities may have to involve development of audit-specific tools or further development of existing materials and tools. In such cases, it requires a preliminary review of the facility or activities to map out specific details and requirements to be observed in the development of the particular tools (Young 1994).

2. *A systematic scrutiny or review of a facility:* Depending on the orientation and goals of audit, the focus of the scrutiny differs. For example, an environmental audit can be conducted without dealing with the processes involved in production. However, an audit, which is considered part of a program of preventive environmental protection, needs to deal with the production processes and material flows. Common to all audits is an analysis and evaluation of the information that has been obtained and an analysis of the outcome vis-à-vis the goals and expectations of the initiator and it helps to identify the areas of improvement.

3. *Reporting:* This step involves the reporting of observation of deficiencies and possible alternatives. It is important to be aware that an environmental audit by itself does not solve any problems. In fact, during the work on the audit, it may appear that the environmental problem is increasing because the audit process brings to attention the hitherto unknown problems or deficiencies. Audits often point to the need for the changes in organization and improvements in education, increased environmental responsibility, and investments in new equipment and environmental protection technology. An important precondition for the success of an audit, especially if it is internally initiated, is that everyone is prepared to accept the consequences and take steps to solve the deficiencies and problems, which the audit may reveal. Thus, the persons whose areas of responsibility have been the object of scrutiny must make decisions and have plans in place to eliminate the problems.

4. *Follow-up:* Following up of the results is an important part of the audit process. An evaluation of the results of the remedial actions is a logical step and can be done either as part of the subsequent audit or as part of a continuing process of enhancing environmental protection procedures. However, a mere follow up on the results may result in a control routine and a stagnation in work on environmental problems. It is, therefore, important that a follow-up process also analyzes the methodology and orientation of the previous audits to ensure that it becomes a dynamic factor.

Types of Environmental Audit

There are two main types of environmental audits, i.e., objective-based and client-driven.

Objective-Based Audit Types

As mentioned earlier, environmental audit covers assessment of any activity that impinges on the environment. The scope and objectives of the audit more usefully distinguish different audit categories and how the audit results are to be used. However, it must be noted that the objectives and scope are often a combination of several audit types and are usually defined on a case-by-case basis (Table 3.4). Organizations develop audit programs to fit their own particular needs. Based on objectives, environmental audits can be categorized as under:

1. *Liabilities audit:* Compliance audit, operational risk audit, acquisition audit, and health and safety audit form liabilities audit. These are often conducted as a prelude to gaining insurance cover and as a means of demonstrating the regulatory compliance. Compliance audit, probably the most common form of environmental audit is a verification process, whereby the facility establishes the extent to which it complies with environmental legislations, regulations, and emission limits. Operational risk concentrates on the potential frequency and consequences of environmentally damaging activities in the raw material and product storage/handling and manufacturing process. Compliance with

Table 3.4 Objectives-based audit types

Environmental audit categories	Environmental audit types	
Liabilities audits	Management audits	Activities audits
Compliance audit	Corporate audit	Site audit
Operational risk audit	Systems audit	Waste audit
Acquisition audit	Policy audit	Product audit
Health & safety audit	Issues audit	Cross-boundary audit

regulations does not necessarily reduce liability due to operational risks. Acquisition audits assess the liability due to contaminated land and building remediation costs. Health and safety audits normally form part of health, safety, and environment (HSE) audit and involve assessment of adequacy of personal protective equipments (e.g., safety shoes, goggles, helmets, etc.), emergency preparedness and disaster management plans (Hillary 1998).

2. *Management audit:* Corporate audit, system audit, policy audit, and issues audit form management audit. They pay considerable attention to management systems as they guide the efficient and effective running of the operations. A corporate audit is initiated by the main Board of a parent company and is concerned with organization structure, roles and responsibilities, policy implementation, awareness and communications. This is carried out as a reassurance to the main Board that its aims and objectives are being implemented throughout the corporate structure. Management system audits are carried out to check the systems against the policy and standards such as British Standard 7750 or ISO 14001. Policy audit is carried out to review and reassess the relevance of a policy in light of developments (legal, technical, financial) within the organization and outside. Issues audit is carried out to establish environmental management plan and targets.

3. *Activities audit:* Site audit, waste audit, product audit, and cross-boundary audit form activities audit. These cover auditing of selected technical and management issues. Environmental site audit examines all aspects of the facilities performances with respect to the environment. It combines most of the elements of other types of EA and, when undertaken in depth, involves considerable time and cost. The waste audits are of two types. The first identifies and quantifies waste streams and is a precursor to waste minimization programs. The second type assesses waste management practices and procedures. Product audits cover several aspects of their environmental impacts through design, manufacture, use, and disposal. Such audits are a prerequisite for identifying environmentally friendly products for "Green Labeling." Cross-boundary audits assess activities, which cut across departments or business units (e.g., transport and supply chain audits).

Client-Driven Audit Types

The different types of audits based on the client, who has commissioned or ordered the audit procedure are (Table 3.5):

Table 3.5 Client-driven types of environmental audit

Category of audit	Ordered by	Desired result
Regulatory external	Regulatory authority	Enhanced oversight
Independent external	Buyer, bank, customer, insurance firm, etc.	Objective information
Internal	Top management, members of board	Reduced risk
Third party	Top management, members of board	Certified environmental protection system

1. *Regulatory external audit:* This often entails an examination carried out by or for an environmental regulatory agency, with the goal of ensuring that a facility is meeting the relevant legislation and regulations. The regulatory agency can use the methodology of audit as a tool to systematically enhance its overview, including the possibility of verifying the accuracy of any reports, which a company is required to submit to the authority.
2. *Independent external audit:* This is conducted by external auditors entitled to perform audits. As the environmental factors have gained importance for a firm's market relations, shareholders such as banks and investment funds, insurance companies, environmental groups, potential buyers, customers, local government and environmentally aware citizens are demanding independent external audits to assess how the firm deals with environmental issues.
3. *Internal environmental audit:* This often involves an inquiry commissioned by management. In practice, such audits are commonly ordered by senior management located at some distance, in both physical and operational senses, from the factory or site of environmental concern. In such cases, the environmental audits are internal in that the results remain within the organization. However, for the facility under investigation, the internal audit has the same effect as an external audit. One reason why firms conduct internal environmental audits is to diminish their liability to pay fines, damages or cleanup costs as a result of breaking the laws (e.g., releasing more emissions than permitted).
4. *Third-party audits:* These represent the audits carried out by the organizations to verify as to whether internal/external audits meet the set standards or not.

General Audit Methodology

The focus of environmental audit differs, depending on the specific requirements of the clients. The audit methods and tools are tailored to suit the purpose for which they are intended (Fig. 3.9). However, any audit program must conform to the following basic structural framework (Barton and Bruder 1995):

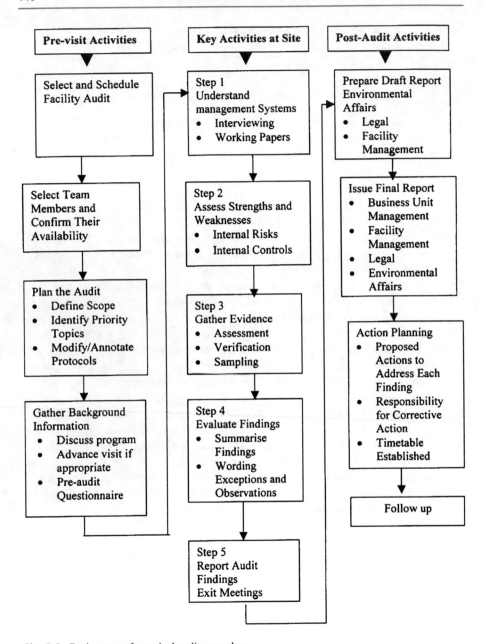

Fig. 3.9 Basic steps of a typical audit procedure

1. Explicit top management support for environmental audit and commitment to follow-up on audit findings
2. An environmental audit function independent of audited activities

3. Adequate team staffing and auditor training
4. Explicit audit program objectives, scope, resources, and frequency
5. A process, which collects, analyzes, interprets, and documents information sufficient to achieve audit objectives
6. A process, which includes specific procedures to promptly prepare candid, clear, and appropriate written reports on audit findings, corrective measures and schedules for implementation
7. A process, which includes quality assurance procedures to assure the accuracy and thoroughness of environmental audits

References

Abaza, H. (2000). Strengthening future environmental assessment practice: An international perspective. In *Environmental assessment in developing and transitional countries*. New York: Wiley.

Abaza, H., Bisset, R., & Sadler, B. (2004). *Environmental impact assessment and strategic environmental assessment: Towards an integrated approach*, United Nations Environment Program.

Ahmad Y. J., & Sammy G. K. (1987). *Guidelines to environmental impact assessment in developing countries*, UNEP Regional Seas Reports and Studies No. 85, UNEP.

Anil, A. (1982). *The state of India's environment* (42 p). New Delhi: Centre for Science and Environment.

Annandale, D. (2001). Developing and evaluating environmental impact assessment systems for small developing countries. *Impact Assessment and Project Appraisal, 19*(3), 187–193.

Arts, J. (2008). Chapter 18 – The importance of EIA follow-up. In T. B. Fischer, P. Gazzola, U. JhaThakur, I. Belcakova, & R. Aschemann (Eds.), *Environmental assessment lecturers' handbook* (pp. 183–196). Bratislava: ROAD Bratislava.

Arts, J., & Morrison, S. A. (2004). Theoretical perspectives on EIA and follow-up. In A. Morrison-Saunders & J. Arts (Eds.), *Assessing impact: Handbook of EIA and SEA follow-up* (pp. 22–41). London: Earthscan.

Barton, H., & Bruder, N. (1995). *A guide to local environmental auditing*. London: Earthscan.

Canter, L. W. (1996). *Environmental impact assessment*. New York: McGraw-Hill.

Canter, L. W. (1977). *Environmental Impact Assessment* (331 pp). New York: McGraw-Hill.

Clark, B. D., Chapman, K., Bisset, R., & Wathern, P. (1978). Methods of environmental impact analysis. *Built Environment, 4*(2), 111–121.

Donnelly, A., Jones, M. B., O'Mahony, T., & Byrne, G. (2006). Decision support framework for establishing objectives, targets and indicators for use in SEA. *Impact Assessment and Project Appraisal, 24*, 151–157.

Eccleston, C. H. (2008). *NEPA and environmental planning: Tools, techniques, and approaches for practitioners* (pp. 148–149). Boca Raton, FL: CRC Press.

Eccleston, C. H. (2014). *The EIS Book: Managing and preparing environmental impact statements*. Chapter 6. Boca Raton, FL: CRC Press.

European Commission, (EC). (1999). *Guidelines for the assessment of indirect and cumulative impacts as well as impact interactions* (172 pp). Luxembourg: Office for Official Publications of the European Communities.

Glasson, J., Therival, R., & Chadwick, A. (1994). *Introduction to environmental impact assessment*. London: UCL Press.

Gruenewald, D. A. (2004). A Foucauldian analysis of environmental education: Toward the socioecological challenge of the Earth Charter. *Curriculum Inquiry, 34*(1), 71–107.

Hillary, R. (1998). Environmental auditing: Concepts, methods and developments. *International Journal of Auditing, 2,* 71–85.

Johnson, E. W., & Fricke, S. (2011). Ecological threat and the founding of U.S. National Environmental Movement Organizations. *Social Problems, 58*(3), 305–309.

Lee, M., & Abbot, C. (2003). The usual suspects? Public participation under the Aarhus Convention'66. *MLR, 80,* 82–85.

Leopold, L. B., Clarke, F. E., Hanshaw, B. B., & Balsley, J. S. R. (1971). *A procedure for evaluating environmental impact. Geological Survey Circular 645.* Washington: U.S. Geological Survey.

Lieberman, G. A. (2013). *Education and the environment: Creating standards-based programs in schools and districts.* Cambridge, MA: Harvard Education Press.

Lyer, V. G. (2017). Strategic environmental assessment (SEA) process for sustainable mining and mineral management development. *Open Access Library Journal, 4,* 2–28.

Malone, K. (1999). Environmental education researchers as environmental activists. *Environmental Education Research, 5*(2), 163–177.

Mehta, M. G. (2010). A river of no dissent: Narmada Movement and coercive Gujarati nativism. *South Asian History and Culture, 1*(4), 509–528.

Morris, P., & Therivel, R. (Eds.). (2001). *Methods of environmental impact assessment* (2nd ed.). London: Spon Press.

Naber, H. (2012). *Environmental impact assessment.* Washington, DC: World Bank.

Ogola, A. (2007). *Environmental impact assessment general procedures.* Lake Naivasha: UNU-GTP and KenGen.

Pendse, Y. D., Rao, R. V., & Sharma, P. K. (1989). Environmental impact assessment methodologies. *International Journal of Water Resources Development, 5*(4), 252–258.

Petsonk, C. A. (1990). The role of the United Nations Environment Programme (UNEP) in the development of International Environmental Law. *American University International Law Review, 5*(2), 351–391.

Prasad, B. C. (1980). *Ecosystem of the Central Himalaya* (32 pp). Gopeshwar: Dasholi Gram Swarajya Mandal.

Smyth, J. C. (2006). Environment and education: A view of a changing scene. *Environmental Education Research, 12*(3–4), 247–264.

Stapp, W. B. (1971). An environmental education program (K-12), based on environmental encounters. *Environment and Behaviour, 3,* 263–283.

Stewart, J. H. (2001). Challenges for environmental education: Issues and ideas for the 21st century. *BioScience, 51*(4), 283.

UNECE. (1998). *UN Environment Principle 10: Bali guidelines for the development of National Legislation on Access to Information, Public Participation and Access to Justice in environmental Matters.* In International Conference Centre Geneva (CICG), Interpretation in English and Russian, p. 4.

UNESCO/UNEP. (1977). *Intergovernmental conference on environmental education.* Paris: Tbilisi Major Environmental Problems in Contemporary Society.

Westman, W. E. (1985). *Ecology, impact assessment and environmental planning* (532 pp). New York: Wiley.

World Business Council for Sustainable Development, (WBCSD). (2005). *Environmental and social impact assessment (ESIA) guidelines* (54 p).

Young, S. S. (1994). *Environmental auditing.* Des Plaines: Cahners Publishing Company.

Environmental Acts and Legislation

4

Abstract

Although, the environment has been the mother of all resources it has been at the receiving when it comes to its interaction with man. Man for his own benefits and by his own acts has almost changed its entire outlook. Pursuing the anthropocentric philosophy of life man has further created an imbalance in its natural equilibrium which brought him to a point where he started thinking about the natural balance and his own role in disturbing the same. Thus, man started redefining the constitutional book of his life by bringing in a series of acts and legislations like fundamental duties, state policies, national and state laws, common and statutory law remedies for either reverting the damage done or at least protecting whatever was left undisturbed for his own sake and for the sake of his future generation. These legislative provisions have doubtlessly played a role and are continuing to do so for the protection of environment.

Keywords

Environmental acts · Directive principles · Law remedies · Statutory law remedies · National laws

Environmental concern in India is as old as the Indian civilization and traditionally this concern was mostly aimed at conservation of forests, wildlife, and other natural resources. But after independence the issues of main concern shifted to sanitation, public health, and hygienic disposal of community wastes as these issues resulted in many outbreaks of deadly communicable diseases (Iyer 1984). However, now the concern has broadened to include sustainable development. The period from independence to 1972 was a period of continuous deterioration and degradation due to two reasons:

(1) Large-scale activities like industrialization, urbanization, rapid population growth, poverty, agricultural intensification, deforestation, loss of wildlife, and pollution.
(2) No serious concern was given to these issues, but the only concern was given to public health and sanitation.

In India, the issue of environmental protection was addressed for the first time in the 4th five year plan (1969–1974). However, it was very significant in the history of environmental conservation because this year 24th General Assembly of United Nations decided to convene an international conference on Environment and Development and a national committee on Environmental planning and coordination was also formed. The conference was held in Stockholm and was attended by the then prime minister of India, Smt. Indira Gandhi. By her participation in the said conference there was a paradigm shift in the concept and concern for environmental protection and conservation (Weinberg and Reilly 2013). Innumerable steps were taken at every level of social, cultural, political, administrative, governmental, and nongovernmental organization for the protection of environment.

Environmental Legislations

Environment legislations consist of all the legal guidelines that are intended to protect our environment. The primary objective of these legislations is to prevent the environment from any kind of damage. Here the objective of environmental protection is achieved by a strict code of conduct, framed with a blend of preventive, promotional and mitigative measures. In general there are three types of legislations relating to environment viz. planning, protective, and preventive legislation. But in most cases legislations are created for protective and preventive purposes. However, planning legislation for environmental quality improvement is almost lacking (Rodgers 1977).

Environmental Legislation in India

Environmental legislation is not a new idea for India but it was present here even in the pre-independence period. Bengal Smoke Nuisance Act–1905 is perhaps one of the oldest environmental protection acts in India followed by the Factories Act, 1948. Similarly, Mines and Minerals (Regulation & Development) Act 1957 is another important environmental legislation formulated in post-independence period. Even before India's independence in 1947, several environmental legislations existed, but the real impetus for bringing about a well-developed framework came only after the United Nations Conference on Human Environment as a series of acts were passed by the parliament thereafter (Leelakhishnan 2002). Under the influence of this declaration, the National Council for Environmental Policy and Planning, within the Department of Science and Technology was set up in 1972, which later on evolved into a full-fledged Ministry of Environment and Forests (MoEF) in 1985. This ministry as on date now named as Ministry of Environment Forest and Climate Change acts as the apex administrative body in the country for regulating and ensuring environmental protection. After the Stockholm Conference, in 1976 constitutional sanction was also given to environmental concerns through the 42nd Amendment, which incorporated them into the Directive Principles of State Policy and Fundamental Rights and Duties (Jaiswal and Jaswal 2003). Since 1970s, an extensive network of environmental legislations has grown in the country. The MoEF and the pollution control boards (both Central and State boards) together form the regulatory and administrative core of the sector (Leelakhishnan 2008).

A policy framework has also been developed to complement the legislative provisions. The Policy Statement for Abatement of Pollution and the National Conservation Strategy and Policy Statement on environment and development were brought out by the MoEF in 1992, to develop and promote initiatives for the protection and improvement of the environment. The EAP (Environmental Action Programme) was formulated in 1993 with the objective of improving environmental services and integrating environmental considerations into developmental programs. Other measures like the sector-specific processes have also been taken by the government to protect and preserve the environment. India has also adopted a federal system in which there is a power sharing agreement between the center and the states. Part XI of the constitution from Articles 245–263 regulates the legislative and administrative relations between the center and states regarding the issues of environment (Leelakhishnan 2002).

- *Article 245*—This article empowers parliament to make laws for the whole or any part of the territory of India, and state legislatures to make laws for their respective states or any part of the state.
- *Article 246*—This article divides various areas of environment and legislation there upon between the union and the state. It provides that if a conflict arrives on any item between the two, the central law shall prevail. It divides them into three lists—the union list, the state list, and the concurrent list.
  - *Union list:* The union list is composed of 97 subjects including industries, mineral resources, and atomic energy over which only the central government has exclusive powers of legislation. It includes industries, mineral resources, atomic energy etc.
  - *State list:* This list comprises of 66 subjects over which only the state governments have the exclusive powers of legislation, subject to their territorial limitation. It includes the areas like public health, sanitation, hospitals, water supply, drainage, and fisheries etc.
  - *Concurrent list:* Concurrent list comprises of 47 subjects over which both center and state legislatures have powers of legislation. The areas present in this list include forests, population control, family planning, and protection of wild animals and birds etc.

Thus the distribution of legislative powers between center and state under federal system of government provides enough provisions to make laws dealing with various environmental problems.

The legal regimes for environmental protection available in India broadly comprise of:

- Common law remedies
- Stationery law remedies

Common Law Remedies

Common law is the law that exists independently of any legislation. It is a set of laws and principles which have been continually developed by courts over hundreds of years. It is important to realize that "common law" is not a fixed or absolute set of written rules in the same sense as statutory or legislatively enacted law (Kumari 1984). There are no specific common law actions designed to protect the environment, as the common law has principally developed to protect the individual's rights and private property rights. However, when an environmental impact also interferes with an individual's right or a private property right, the common law can be used to protect the environment indirectly. Common law remedies against environmental pollution are available under law of torts and it is among the oldest legal remedies to abate pollution (Nayak 1999). Tort is basically a civil wrong other than a breach of trust and a breach of the common law is said to give rise to a "cause of action." Some common law remedies that might be used to protect the environment are:

- Trespass Nuisance
- Negligence Strict Liability

Each of these causes of action can be used to protect different rights in different situations, although it is often the case that several causes of action may be applicable to a single issue. The common law may be used to protect the environment because it may provide a legal recourse that is not otherwise available under the legislation, or because in some situations it may provide remedies that are more desirable or suitable than those available under legislation.

Trespass

Trespass occurs where a person directly, intentionally or negligently and without permission causes some physical interference with another person's property (Kramer 2006). Trespass does not require proof of damage or harm. An example of trespass in an environmental situation might be if a person deliberately sprays pesticides or dumps waste on your property.

Trespass to land is an unlawful or forcible entry on another's realty. In an action for trespass to land, entry upon another's land need not be in person but by causing or permitting a thing to cross the boundary of the premises as well. It may be committed by casting material upon another's land, by discharging water, soot or carbon, by allowing gas or oil to flow underground into someone else's land, but not by mere vibrations or light which are generally classed as nuisances. It is the type of trespasses actions that is generally used in pollution control cases (Kiss and Shelton 2004).

In the case of *Martin* vs. *Reynolds Metal Co.* the deposit on Martin's property of microscopic fluoride compounds, which were emitted in vapor form from the Reynolds' plant, was held to be an invasion of this property—and so a trespass (Kumari 1984).

The line between trespass and nuisance is sometimes difficult to determine. However, the distinction which is now accepted is that trespass is an invasion of the plaintiff's interest in the exclusive possession of his land, while nuisance is an interference with his use and enjoyment of it.

Nuisance

Anything that annoys hurts or interfaces with the quality of life is referred to as nuisance. It can be either in the form of water pollution, dust, smoke, and other hazardous wastes. Dealt under public Nuisance Act, it is the most frequently used common law action. Nuisance law traditionally protects the rights of a landowner to use and enjoy the property. Nuisance is defined as "that activity which arises from the unreasonable, unwarrantable or unlawful use by a person of his own property, working an obstruction or injury to the right of another or to the public, and producing such material annoyance, inconvenience, and discomfort that the law will presume resulting damage" (Rodgers 1982).

Nuisance actions come in two forms: public and private. However, under both private and public nuisance laws, the plaintiff must prove that the defendant's activity *unreasonably interfered* with the use or enjoyment of a protected interest and caused the plaintiff *substantial* harm (Kumari 1984).

Private nuisance is committed where one person (the defendant) substantially and unreasonably interferes with another persons (the plaintiff) right to the use and enjoyment of their land. Unlike trespass, interference can amount to a private nuisance even if it is not direct or intentional. Private nuisance is one of the most commonly used actions to address environmental concerns. "Public nuisance" occurs when a person causes a nuisance which "endangers the life, health, property, morals or comfort of the public" or "obstructs the public in the exercise or enjoyment of rights common to all".

Village of Wilsonville vs. *SCA Services, Inc.* is a case in which the plaintiffs, a village and other governmental bodies, alleged that the defendant's hazardous chemical landfill was a *public* nuisance. The plaintiffs sought to enjoin the operations of the landfill and require removal of toxic waste and contaminated soil. The court found that there was a substantial danger of groundwater contamination and explosions from chemical reactions. Although the damages were *prospective,* the nuisance already was present. Therefore, the court granted an injunction and ordered a site cleanup.

Negligence

When there is a duty to take care of and it has not been taken into consideration, the resulting harm to another person is called as negligence. "Negligence" is the omission to do something which a reasonable man, guided by those ordinary considerations which ordinarily regulate human affairs, would do, or the doing of something which a reasonable and prudent man would not do. Negligence is that part of the law of torts which deals with acts not intended to inflict injury. This is often defined as the "reasonable mans" rule, what a reasonable person would do under all the circumstances.

Negligence may also be used as a cause of action to address environmental harm. To plead negligence, the person bringing the action (the plaintiff) must be able to prove that (Kumari 1984):

1. The defendant owed the plaintiff a "duty of care."
2. The defendant breached this duty.
3. This breach of duty caused damage to the plaintiff.

Nissan Motor Corporation vs. Maryland Shipbuilding and Dry Dock Company: It exemplifies a negligence action in an environmental case. The shipbuilding company's employees failed to follow company regulations when painting ships, allowing spray paint to be carried by the wind onto Nissan's cars. The shipbuilders had knowledge of the likely danger of spray painting, yet failed to exercise due care in conducting the painting operations in question. Hence, the failure amounted to negligence.

Strict Liability

If a person for his/her own purpose bring to his/her own land, collects and keeps what which is likely to be harmful to another, he/she is to compensate for the damage done. It ensures that the burden of damage is to be beard by the industry concerned. After Bhopal Gas tragedy it was replaced by "Absolute liability" articulated by the Supreme Court and adopted by the parliament.

Statutory Law Remedies

These are the laws of national and regional legislatures and local governments. Different activities harmful to environment can be controlled through the legislative framework which includes:

(1) *Indian penal code, (IPC) 1860:* It is a code of conduct of Indian constitution to eradicate different types of wrongs from the society. Its chapter XIV comprises Sections 268–294A in which offences affecting the public health, safety, convenience, decency, and morals are given in Table 4.1.
(2) *Criminal procedure code (CrPC), 1973:* It is another strict code of conduct involved to prevent any type of pollution. Its chapter X Part B contains sections

Table 4.1 Offences affecting the public health, safety, convenience, decency and morals (Kumari 1984)

Chapter XIV: of offences affecting the public health, safety, convenience, decency, and morals	
Article 268	Public nuisance
Article 277	Fouling water of public spring or reservoir
Article 278	Making atmosphere noxious to health
Article 283	Danger or obstruction in public way or line of navigation
Article 284	Negligent conduct with respect to poisonous substance
Article 285	Negligent conduct with respect to fire or combustible matter
Article 286	Negligent conduct with respect to explosive substance
Article 287	Negligent conduct with respect to machinery
Article 289	Negligent conduct with respect to animal
Article 290	Punishment for public nuisance in cases not otherwise provided for
Article 291	Continuance of nuisance after injunction to discontinue

Table 4.2 Chapter X, Part B and C

Chapter X Part B—Public Nuisances	
Section 133	Conditional order for removal of nuisance
Section 134	Service or notification of order
Section 135	Person to whom order is addressed to obey or show cause
Section 136	Consequences of his failing to do so
Section 137	Procedure where existence of public right is denied
Section 138	Procedure where he appears to show cause
Section 139	Power of magistrate to direct local investigation and examination of an expert
Section 140	Power of magistrate to furnish written instructions, etc.
Section 141	Procedure on order being made absolute and consequences of disobedience
Section 142	Injunction pending inquiry
Section 143	Magistrate may prohibit repetition or continuance of public nuisance
Chapter X Part C—Urgent cases of nuisance or apprehended danger	
Section 144	Power to issue order in urgent cases of nuisance or apprehended danger

133–143 and Part C has the directions for speedy and effective remedy against pollution problems and nuisance under section 133 CrPC (Table 4.2). If a person fails to comply with its provision, he/she is prosecuted under section 183 of IPC. It is also used against all statutory bodies like municipal corporation and government bodies etc. Thus both IPC and CrPC, although of ancient vintage, are the contemporary remedial weapons (Juergensmeyer 1971).

Constitutional Perspectives

Protection of environment prior to 42nd amendment was availed through Section 21 of the constitution, stating that "no person shall be deprived of his/her life or personal liability except according to procedures established by law," i.e., ensuring that every persons has fundamental right to life that comprises right to live in a healthy environment.

Fundamental Duties

The Constitution's 42nd Amendment Act, 1976 added a new Part IV A, dealing with "fundamental duties" in the constitution of India. Article 51A (g) specifically deals with the fundamental duties with respect to the environment and provides:

It shall be the duty of every citizen of India to protect and improve the natural environment including forests, rivers and wildlife and to have compassion for living creatures.

Taking protection of environment is a matter of constitutional priority. The fundamental duties are intended to promote peoples participation in restructuring and building a welfare society. Environmental problem is the concern of every citizen and neglect of it is an invitation of disaster. Article 51A (g) gives the right to the citizens to move to court to see that the state performs its duties faithfully in accordance with the law of the land or not (Jaiswal and Jaswal 2003).

The true scope of Article 51A (g) has been best explained by Rajasthan high court in *L.K. Koolwal* vs. *State of Rajasthan* in 1988, a case where in the municipal authority was charged with the "Primary duty" to clean public streets, sewers and all spaces, not being private property which are open to the enjoyment of public, removing of noxious vegetation and all public nuisance and to remove filth, rubbish, night soil, odor or any other noxious or offensive matter. But it failed to perform its "Primary duty" resulting in acute sanitation problems in Jaipur posing a threat to human life. Mr. Koolwal moved the High court on the basis of the right given to it by Article 51A (g) for the enforcement of the duty cast on state.

Directive Principles of State Policy

Part IV of the constitution deals with the directive principles of the state policy. The constitution's (42nd Amendment) Act added a new directive principle in Article 48A dealing with the improvement of environment which provides:

The state shall endeavor to protect and improve the environment and to safeguard the forests and wildlife of the country.

The state cannot treat the obligation of protecting and improving the environment as mere pious obligations. The directive principles are not mere showpieces in the window dressing. They are *"fundamental in the governance of the country"* and they being part of the supreme law of land have to be implemented in letter and spirit. This incorporation of putting obligations on the *"state"* as well as *"citizens"* to protect and improve the environment in the *"Suprema Lax"* is certainly a positive development of Indian law (Jaiswal and Jaswal 2003).

In *Shri Sachindanand pandey* vs. *State of West Bengal* the supreme court pointed out that whenever a problem of ecology is brought before the court, the court is bound to bear in mind Article 48A and 51A(g) of the constitution.

State Laws

The different laws enacted by the states from time to time, to tackle the problems of environment like water and air pollution are explained as under.

The Water (Prevention and Control of Pollution) Act, 1974

Water is a state subject and as such central government cannot make laws relating to it, provided the state legislature authorizes it to do so under clause (I) of Article 252 of the constitution. After seeking the consent of 12 states, parliament passed a legislation regarding water called as Water (Prevention and Control of Pollution) Act. *"An Act to provide for prevention and control of water pollution and the maintaining or restoring of wholesomeness of water, for the establishment, with a view to carrying out the purposes aforesaid, of Boards for the prevention and control of water pollution, for conferring on and assigning to such Boards powers and functions relating thereto and for matters connected therewith"* (Kumar 1992).

This Act represented India's first attempt to comprehensively deal with environmental issues. The Act prohibits the discharge of pollutants into water bodies beyond a given standard, and lays down penalties for noncompliance. The Act was amended in 1988 to conform closely to the provisions of the EPA, 1986. It set up the CPCB (Central Pollution Control Board) to lay down standards for the prevention and control of water pollution at central level. The SPCBs (State Pollution Control Boards) function as per the direction of CPCB and the state government at the state level.

Objectives of the Act

1. Prevention and control of water pollution
2. Sustaining or restoring the wholesomeness of water
3. Establishment of boards for prevention and control of water pollution

Functions of Central Pollution Control Board: The functions of the Central Pollution Control Board as given by Section 16-A of the act are:

(i) To promote cleanliness of streams and wells in different areas
(ii) To advise the central government on various issues of pollution
(iii) To coordinate the work of state boards and to resolve the disputes among them
(iv) To provide technical assistance and support to state pollution control boards
(v) To establish laboratories for water analysis
(vi) To lay down standards for steams and wells

Functions of State Pollution Control Boards: The functions of the State Pollution Control Boards as given by Section 7B of the act are:

(i) Planning a comprehensive program for prevention, control, and abatement of water pollution
(ii) To advise state governments on various issues of pollution
(iii) To collaborate with center pollution control board for handling the issues of pollution
(iv) To evolve economical and reliable methods for disposal and treatment of waste water
(v) To establish and recognize laboratories of water analysis
(vi) To lay down standards for various activities
(vii) To conducting research work for prevention and control of water pollution

Powers of the State Boards

(i) Power to obtain information
(ii) Power to take samples
(iii) Power to obtain a report of the result of analysis by a recognized laboratory
(iv) Power of entry and inspection for performing the entire duty
(v) Power of prohibition on disposal of polluting matter into a stream or well

Besides, the act provides for a permit system, or consent procedure to prevent and control water pollution. It prohibits the disposal of wastes into the water bodies in excess of the standards established by the boards. It also provides that if a person is interested to establish any industry he/she shall get consent from the state and if a person fails to do so the board may order closure, prohibition or regulation of any industry. The boards are also empowered to stop or regulate electric supply, water supply, and other facilities to the unit.

Penalties

Whoever fails to comply with any direction given under Subsection (2) or Subsection (3) of Section 20 within such time as may be specified in the direction shall, on conviction, be punishable with imprisonment for a term which may extend to 3 months or with fine which may extend to 10,000 rupees or with both and in case the failure continues, with an additional fine which may extend to 5000 rupees per day during which such failure continues after the conviction for the first such failure (Jaiswal and Jaswal 2003).

Whoever:

(a) Destroys, pulls down, removes, injures, or defaces any pillar, post, or stake fixed in the ground or any notice or other matter put up, inscribed or placed, by or under the authority of the board

(b) Obstructs any person acting under the orders or directions of the board from exercising his powers and performing his functions under this Act

(c) Damages any work or property belonging to the board

(d) Fails to furnish to any officer or other employees of the board any information required by him for the purpose of this Act

(e) Fails to intimate the occurrence of an accident or other unforeseen acts or events under section 31 to the board and other authorities or agencies as required by that section

(f) Fails in giving any information which he is required to give under this Act

(g) For the purpose of obtaining any consent under section 25 or section 26, knowingly or willfully makes a statement which is false in any material, shall be punishable with imprisonment for a term which may extend to three months or with a fine which may extend to 10,000 rupees or with both

The Air (Prevention and Control of Pollution) Act, 1981

This act was passed under Article 253 of Indian constitution in pursuance of the decisions of Stockholm conference. To counter the problems associated with air pollution, ambient air quality standards were established, under this Act. The Act which provides means for the control and abatement of air pollution seeks to combat air pollution by prohibiting the use of polluting fuels and substances, as well as by regulating appliances that give rise to air pollution. Under the Act establishing or operating of any industrial plant in the pollution control area requires consent from state boards. The boards are also expected to test the air in air pollution control areas, inspect pollution control equipment, and manufacturing processes (Bakshi 1993).

Objectives of the Act
The objectives of this act are:

1. To provide for the prevention, control, and abatement of air pollution in order to preserve the quality of air

2. To provide powers to the boards to take a deterrent action on the accused and performing certain functions necessary to improve the environment

Hence in the central board and state board constituted under Section 3 of the Water (Prevention and Control of Pollution) Act, 1974 (6 of 1974) were given the consent that they, shall, without prejudice to the exercise and performance of its powers and functions, for the Prevention and Control of Air Pollution under this Act.

Powers and Functions of Boards

Subject to the provisions of this Act, and without prejudice to the performance, of its functions under the Water (Prevention and Control of Pollution) Act, 1974 the main functions of the central board shall be to improve the quality of air and to prevent, control, or abate air pollution in the country (Bakshi 1993).

The central board may:

(a) Advise the central government on any matter concerning the improvement of quality of air and the prevention, control, or abatement of air pollution
(b) Plan and cause to execute a nation-wide program for the prevention, control, or abatement of air pollution
(c) Coordinate the activity of the states and resolve disputes among them
(d) Provide technical assistance and guidance to the state boards, carry out and sponsor investigations and research relating to problems of air pollution
(e) Plan and organize the training of persons engaged or to be engaged in programs for the prevention, control, or abatement of air pollution on such terms and conditions as the central board may specify
(f) Organize through mass media a comprehensive program regarding the prevention, control, or abatement of air pollution
(g) Collect, compile, and publish technical and statistical data relating to air pollution and the measures devised for its effective prevention, control or abatement and prepare manuals, codes or guides relating to prevention, control, or abatement of air pollution
(h) Lay down standards for the air quality
(i) Collect and disseminate information in respect of matters relating to air pollution
(j) Establish or recognize a laboratory or laboratories to enable the central board to perform its functions under this section efficiently
(k) Delegate any of its functions under this Act, generally or specially to any of the committees appointed by it

The functions of a State Board shall be:

(a) To plan a comprehensive program for the prevention, control, or abatement of air pollution and to secure the execution thereof
(b) To advise the state government on any matter concerning the prevention, control, or abatement of air pollution
(c) To collect and disseminate information relating to air pollution
(d) To collaborate with the central board in organizing the training of persons engaged or to be engaged in programs relating to prevention, control, or abatement of air pollution and to organize mass-education programs relating thereto
(e) To inspect, at all reasonable times, any control equipment, industrial plant or manufacturing process and to give, by order, such directions to such persons as

it may consider necessary to take steps for the prevention, control, or abatement of air pollution

(f) To inspect air pollution control areas at such intervals as it may deem necessary, assess the quality of air therein and take steps for the prevention, control, or abatement of air pollution in such areas

(g) To lay down (in consultation with the central board) standards for emission of air pollutants into the atmosphere from industrial plants and automobiles or for the discharge of any air pollutant into the atmosphere from any other source whatsoever

(h) To advise the state government with respect to the suitability of any premises or location for carrying out any industrial operation which is likely to cause air pollution

(i) To perform such other functions as may be prescribed or as may, from time to time, be entrusted to it by the central board or the state government

(j) To do such other things and to perform such other acts as it may think necessary for the proper discharge of its functions and generally for the purpose of carrying into effect the purposes of this Act

(k) To establish or recognize a laboratory or laboratories to enable the State Board to perform its functions under this section efficiently (Bakshi 1993)

Penalties and Procedure

Whoever fails to comply with the provisions of Section 21 or Section 22 or directions issued under Section 31 A, shall, in respect of each such failure, be punishable with imprisonment for a term which shall not be less than 1 year and 6 months but which may extent to 6 years and with a fine. In case the failure continues, an additional fine which may extend to 5000 rupees per day during which failure continues after the conviction for the first-such failure may be imposed.

If the failure continues beyond a period of 1 year after the date of conviction, the offender shall be punishable with imprisonment for a term which shall not be less than 2 years but which may extend to 7 years, with fine.

Whoever:

(a) Destroys, pulls down, removes, injures, or defaces any pillar, post, or stake fixed in the ground or any notice or other matter put up, inscribed or placed, by or under the authority of the board

(b) Obstructs any person acting under the orders or directions of the board from exercising his powers and performing his functions under this Act

(c) Damages any works or property belonging to the board

(d) Fails to furnish to the board, or any officer or other employee of the board if any information required by the board or such officer or other employee for the purpose of this Act

(e) Fails to intimate the occurrence of the emission of air pollutants into the atmosphere in excess of the standards laid down by the state board or the apprehension of such occurrence, to the state board and other prescribed authorities or agencies as required under Subsection (1) of Section 23

(f) In giving any information which he is required to give under this Act

(g) For the purpose of obtaining any consent under Section 21, makes a statement which is false in any material shall be punishable with imprisonment for a term which may extend to 3 months or with fine which may extend to 10,000 rupees or with both

National Laws

The national laws present in the constitution of India pertaining to the protection and improvement of environment are explained below.

The Environment Protection Act, 1986

This Act (may be called the environment (protection) Act 1986) came into force on November 19, 1986 in the whole of Indian under Article 253 of Indian constitution in pursuance with the declarations of Stockholm conference. It is an umbrella legislation to provide powers and framework for central government to coordinate activities of various central and state authorities established under previous laws. It was promulgated to provide for the protection and improvement of environment and the matters connected there with. Under this Act, the central government is empowered to take measures necessary to protect and improve the quality of the environment by setting standards for emissions and discharges, regulating the location of industries; management of hazardous wastes, and protection of public health and welfare.

Objectives
The objectives of this Act are:

(a) Protection and improvement of environment

(b) Implementation of the decisions of United Nations conference on human environment

(c) Covering the uncovered gap in the areas of environment

(d) Coordination of the work of various agencies

(e) Providing a deterrent punishment to those who endanger the safety of environment and health

(f) Maintenance of harmony between environment and human beings

The Act empowers the central government to take all necessary measures for the protection and improvement of environment like:

- Laying down standards for environmental safety
- Laying down standard for emission or discharge of environmental pollutants
- Laying down procedures and safeguards for prevention of articles
- Laying down procedures for handling hazardous wastes
- Designation of certain nationwide programs for preventing, controlling and abating pollution
- Establishment of environmental laboratories
- Constitution of authority to exercise its powers
- Appointment of officers, their powers and functions
- Preparation of manuals, codes, and guides related to pollution

Power of Central Government

Subject to different provisions of this Act, the central government shall have the powers to take all such measures as it deems necessary or expedient for the purpose of protecting and improving the quality of environment and preventing, controlling, and abating environmental pollution (Sahasranaman 2009). In particular and without prejudice to the generality of provisions of Subsection (1) such measures may include all or any of the matters namely:

1. Coordination of actions by the state government, officers, and other authorities
2. Planning and extension of nation-wide program for the prevention, control, and abatement of environmental pollution
3. Laying down standards for the quality of environment in its various aspects
4. Laying down standards for emission or discharge of environmental pollutants from various sources whatsoever
5. Restrictions of areas in which any industry, operation or process, or class of industries, operations or processes shall not be carried out or shall be carried out subject to certain safeguards
6. Laying down procedures and safeguards for the prevention of accidents which may cause environmental pollution and remedial measures for such accidents
7. Laying down procedures and safeguards for the handling of hazardous substances
8. Examination of such manufacturing processes, materials, and substances as are likely to cause environmental pollution
9. Carrying out and sponsoring investigations and research relating to problems of environmental pollution
10. Inspection of any premises, plants, equipment, machinery, manufacturing or other processes, materials or substances and giving by order, of such direction to such authorities, officers and persons as it may consider necessary to take steps, for prevention, control, and abatement of environmental pollution
11. Establishment or recognition of environmental laboratories and institutes to carry out functions entrusted to such environmental laboratories and institutes under this Act
12. Collection and dissemination of information in respect of matters relating to environmental pollution

13. Preparation of manuals, codes, or guides relating to the prevention control and abatement of environmental pollution
14. Such other matters as the central government deems necessary or expedient for the purpose of securing the effective implementation of the provisions of this Act
15. Powers of entry and inspection—for performing any function entrusted under legislation
16. Power to take samples—(of air, water, soil) from the factory for analysis
17. Power to establish laboratories for analysis purposes
18. Power to give direction:
    - Closure, prohibition, or regulation of any industry, operation, or process; or
    - For stoppage or regulation of the supply of electricity or water or any other service

Penalties
Any person failing in any of the provisions of this Act shall be punishable with imprisonment for a term which may extend up to 5 years or with fine which may extend up to rupees 1 lakh or both. In case the violation or contravention continues beyond a period of 1 year after the date of conviction, the offender shall be punishable with imprisonment for a term of 7 years.

Rules notified under Environment (Protection) Act: Various rules notified under EPA-1986 are explained below.

Biomedical Waste (Management and Handling) Rules, 1998
In exercise of the powers conferred by Section 6, 8 and 25 of the EPA, 1986, the central government notified the rules for the management and handling of biomedical wastes. The rules were published on July 20, 1998 and appeared in the official gazette of India on July 27, 1998. These rules apply to all persons who generate, collect, receive, store, transport, dispose, or handle the biomedical wastes in any form. These rules regulate the disposal of all types of biomedical wastes including blood, body parts, medicines, glass, solid wastes, animal wastes, liquids, and biotechnological wastes. The biomedical waste means any waste which is generated during the diagnosis, treatment, or immunization of human beings or animals or in research activities pertaining thereto or in the production or testing of biological and including categories mentioned in schedule I of the rules (Sahasranaman 2009).

Objective The main objective of these rules is to take steps to ensure safety to health and environment.

Duty of the Occupier It shall be the duty of every occupier of an institution generating biomedical wastes which includes a hospital, nursing home, clinic, dispensary, veterinary, animal house, pathological laboratory, blood bank by whatever name called, to take all steps to ensure that such waste is handled without any adverse effect to human health and environment.

Treatment and Disposal

1. Biomedical waste shall be treated and disposed of in accordance with Schedule I, and in compliance with the standards prescribed in Schedule V.
2. Every occupier, where required, shall set up in accordance with the time schedule in Schedule VI, requisite biomedical waste treatment facilities like incinerator, autoclave, microwave system for the treatment of waste, or, ensure requisite treatment of wastes at a common waste treatment facility or any other waste treatment facility.

Segregation, Packaging, Transportation, and Storage

1. Biomedical waste shall not be mixed with other wastes.
2. Biomedical waste shall be segregated into labeled containers/bags according to Schedule III at the point of generation in accordance with Schedule II prior to its storage, transportation, treatment, and disposal.
3. If a container is transported from the premises where biomedical waste is generated to any waste treatment facility outside the premises, the container shall, apart from the label prescribed in Schedule III, also carry information prescribed in Schedule IV.
4. Notwithstanding anything contained in the Motor Vehicles Act, 1988, or rules there under, untreated biomedical waste shall be transported only in such vehicle as may be authorized for the purpose by the competent authority as specified by the government.
5. No untreated biomedical waste shall be kept stored beyond a period of 48 h.

The pollution control boards and pollution control committees are the prescribed authorities for the implementation of these rules.

Municipal Solid Waste (Management and Handling) Rules, 2000

In exercise of the powers conferred by the EPA, 1986 the Ministry of Environment and Forests, Government of India notified the Municipal solid wastes (Management and handling) Rules 2000 on September 25, 2000, with the aim to take all the necessary steps to properly manage and handle the municipal solid wastes, so as to protect the human health and environment. These rules make the municipal bodies/local bodies responsible for the management and handling of these wastes (Table 4.3). It includes four main schedules.

Schedule I: In this deadline have been given for:

1. Setting up waste processing and disposal facilities
2. Monitoring the performance of these facilities
3. Improving existing and identification of new landfill sites for future use

Table 4.3 Authorities and responsibilities

S. no	Agency/authorities	Responsibility
1	*Municipal authorizes*	(i) Ensuring that municipal solid wastes to be handled as per rules (ii) Seeking authorization from state pollution control board (SPCB) for setting up waste processing and disposal facility including landfills (iii) Furnishing annual report (iv) Complying with different schedule of the rules
2	*State government* Secretary in-charge of Department of Urban development	Overall responsibilities for the enforcement of the provision of the rules in the metropolitan cities. Overall responsibilities for the enforcement of the provision of the rules within the territorial limits of their jurisdiction.
3	*Central pollution control board (CPCB)*	(i) Coordinate with state boards and committees with reference to implementation and review of standards and guidelines and compilation of monitoring data. (ii) Prepare consolidated annual report on management of municipal solid wastes for forwarding it to central government along with its recommendations before the 15th of December every year. (iii) Laying down standards on waste processing/disposal technologies including approval of technology.
4	*State pollution control board (SPCB)*	(i) Monitoring the compliance of the standards regarding ground water, ambient air, leachate quality and the compost quality including incineration standards as specified under schedule II, III, and IV. (ii) Issuance of authorization to the municipal authority of operator of a facility stipulating compliance criteria and standards. (iii) Preparation of an annual report with regard to the implementation of the rules to be submitted to CPCB.

Schedule II: describes the standards for:

1. Segregation
2. Collection
3. Storage
4. Transformation
5. Processing and
6. Disposal of solid wastes

Schedule III: it provides specifications for:

1. Site selection
2. Development and establishment of sanitary landfills
3. Landfill closure and post care, once it is covered

4. Controlling air, water pollution and monitoring standards

Schedule IV: Indicates waste processing options including; standards for composting, treated leachates and incinerations.

Recycled Plastic Manufacturing and Usage Rules, 1999

In exercise of the powers conferred by the EPA–1986, the ministry of Environment and Forests, Government of India notified the plastic manufacturing and usage rules 1999, for regulating the manufacturing and use of recycled plastic carry bags and containers (Jaiswal and Jaswal 2003).

Features of the Rules

1. The usage of carry bags and containers made of recycled plastic, for food item is strictly prohibited.
2. For the manufacture of carry bags and containers, made of plastic, the following conditions should be satisfied:
    (i) Carry bags and containers made of virgin materials should be in natural shade or white.
    (ii) Carry bags and containers made of recycled plastic shall be manufactured using pigment and colorants as per IS 9833: 1981, entitled *"list of pigments and colorants used in plastics in contact with foodstuffs, pharmaceuticals and drinking water."*
    (iii) Recycling shall be undertaken according to Bureau of Indian standards specification: IS 14534: 1998, entitled: *"The guidelines for Recycling of plastics."*
    (iv) Marking and codification of carry bags and containers shall also be as per BIS specification: IS 14534: 1998, entitles *"The Guidelines for recycling of plastics."*
    (v) Manufacturers shall print on each packet of carry bags as to whether these are made of *"recycled material"* or *"virgin material."*
    (vi) Minimum thickness of carry bags shall not be less than 20 microns.
    (vii) The plastic industry association, through their member units, shall undertake self-regulatory measures.

Prescribed Authority

(a) The prescribed authority for enforcement of the provisions of these rules related to manufacture and recycling, shall be the state pollution control boards in respect of states and the pollution control Committees in respect of Union Territories.
(b) The prescribed authority for enforcement of the provisions of these rules related to the use, collection, segregation, transportation, and disposal shall be the District Collector/Deputy Commissioner of the concerned district where no such authority has been constituted by the State Government/Union Territory administration under any law regarding non-biodegradable wastes.

Table 4.4 Ambient air quality standards in respect of noise

Area code	Category of area/zone	Limits in dB(A) Leq[a]	
		Day time	Night time
A	Industrial area	75	70
B	Commercial area	65	55
C	Residential area	55	45
D	Silence zone	50	40

[a]dB(A) Leq denotes the time weighted average of the level of sound in decibels on scale A which is relatable to human hearing
(a) Day time shall mean from 6:00 a.m. to 10:00 p.m.
(b) Night time shall mean from 10:00 p.m. to 6:00 a.m.
(c) Silence zone is an area comprising not less than 100 meters around hospitals, educational institutions, courts, religious places or any other area which is declared as such by the competent authority
(d) Mixed categories of areas may be declared as one of the four abovementioned by the competent authority

Noise Pollution (Regulation and Control) Rules, 2000

Whereas the increasing ambient noise levels in public places from various sources, like industrial activity, construction, fire crackers, sound-producing instruments, generator sets, loudspeakers, public address systems, music systems, vehicular horns and other mechanical devices have deleterious effects on human health and the psychological well-being of the people; it is considered necessary to regulate and control noise producing and generating sources with the objective of maintaining the ambient air quality standards in respect of noise. In exercise of the powers conferred by the EPA, 1986, the central government introduced one more set of rules namely the Noise Pollution (Regulation and Control) Rules 2000 to control noise pollution (Table 4.4).

Objective: The main objectives of the set of rules are:

(i) Regulation and control of noise producing and generating sources
(ii) Maintaining ambient air quality standards with respect to noise

Features: Various features of the Noise Pollution (Regulation and Control) Rules include:

1. The State government shall categorize the areas into silence, residential, commercial, and industrial zones/areas to implement noise standards for them
2. The state government shall take measures for abatement of noise emanating from vehicular movements, blowing of horns, bursting of sound-emitting fire crackers, use of loudspeakers or public address systems and sound-producing instruments and ensure that the existing noise levels do not exceed the ambient air quality standards specified under rules.
3. An area comprising not less than 100 meters around hospitals, educational institutions, and courts may be declared as silence area/zone for the purpose of these rules.

4. The noise levels in any area/zone shall not exceed the ambient air quality standards in respect of noise as specified in the Schedule.

5. The authority shall be responsible for the enforcement of noise pollution control measures and the due compliance of the ambient air quality standards in respect of noise.

6. The respective state pollution control boards or pollution control committees in consultation with the central pollution control board shall collect, compile, and publish technical and statistical data relating to noise pollution and measures devised for its effective prevention, control, and abatement.

7. A loudspeaker or a public address system or any sound-producing instrument or a musical instrument or a sound amplifier shall not be used at night time except in closed premises for communication within, like auditoria, conference rooms, community halls, banquet halls or during a public emergency.

8. The noise level at the boundary of the public place, where loudspeaker or public address system or any other noise source is being used shall not exceed 10 dB (A) above the ambient noise standards for the area or 75 dB (A) whichever is lower;

9. The peripheral noise level of a privately owned sound system or a sound-producing instrument shall not, at the boundary of the private place, exceed by more than 5 dB (A) the ambient noise standards specified for the area in which it is used.

10. No horn shall be used in silence zones or during night time in residential areas except during a public emergency.

11. Sound-emitting fire crackers shall not be burst in silence zones or during night time.

12. Sound-emitting construction equipment shall not be used or operated during night time in residential areas and silence zones.

Consequences of Any Violation in Silence Zones/Areas

Whoever, in any place covered under the silence zone/area commits any of the following offence, shall be liable for penalty under the provisions of the Act:

1. Whoever plays any music or uses any sound amplifiers
2. Whoever beats a drum or tom-tom or blows a horn either musical or pressure, or trumpet or beats or sounds any instrument
3. Whoever exhibits any mimetic, musical or other performances of a nature to attract crowds
4. Whoever bursts sound-emitting fire crackers
5. Whoever uses a loudspeaker or a public address system

Complaints to Be Made to the Authority

1. A person may, if the noise level exceeds the ambient noise standards by 10 dB or more given in the corresponding columns against each area/zone or, if there is a

violation of any provision of these rules regarding restrictions imposed during night time, make a complaint to the authority.

2. The authority shall act on the complaint and take action against the violator in accordance with the provisions of these rules and any other law in force.

Notifications Issued Under EPA, 1986

1. *Doon Valley Notification (1989):* It prohibits the setting up of an industry in which the daily consumption of coal/fuel is more than 24 MT (million tons) per day in the Doon Valley.
2. *Revdanda Creek Notification (1989):* This notification prohibits setting up of industries in the belt around the Revdanda Creek as per the rules laid down in the notification.
3. *Coastal Regulation Zone Notification (1991):* It regulates activities along coastal stretches. As per this notification, dumping ash or any other waste in the CRZ is prohibited. The thermal power plants (only foreshore facilities for transport of raw materials, facilities for intake of cooling water and outfall for discharge of treated waste water/cooling water) require clearance from the MoEF.
4. *Dhanu Taluka Notification (1991):* It is a notification under which the district of DhanuTaluka has been declared as an ecologically fragile region and setting up power plants in its vicinity has been declared prohibited.
5. *The Environmental Impact Assessment of Development Projects Notification, (1994 and as amended in 1997):* As per this notification:
   (a) All projects listed under Schedule I require environmental clearance from the MoEF.
   (b) Projects under the delicenced category of the New Industrial Policy also require clearance from the MoEF.
   (c) All developmental projects whether or not under the Schedule I, if located in fragile regions must obtain MoEF clearance.
   (d) Industrial projects with investments above Rs. 500 million must obtain MoEF clearance and are further required to obtain a LOI (Letter Of Intent) from the Ministry of Industry, and an NOC (No Objection Certificate) from the SPCB and the State Forest Department if the location involves forestland. Once the NOC is obtained, the LOI is converted into an industrial license by the state authority.
   (e) The notification also stipulated procedural requirements for the establishment and operation of new power plants. As per this notification, two-stage clearance for site-specific projects such as pithead thermal power plants and valley projects is required. Site clearance is given in the first stage and final environmental clearance in the second. A public hearing has been made mandatory for projects covered by this notification. This is an important step in providing transparency and a greater role to local communities.
6. *Ash Content Notification (1997):* It required the use of beneficiated coal with ash content not exceeding 34% with effect from June 2001 (the date later was

extended to June 2002). This applies to all thermal power plants located beyond 1000 km from the pithead and any thermal power plant located in an urban area or, sensitive area irrespective of the distance from the pithead except any pithead power plant.

7. *Taj Trapezium Notification (1998):* This notification provided that no power plant could be set up within the geographical limits of the Taj Trapezium assigned by the Taj Trapezium Zone Pollution (Prevention and Control) Authority.

8. *Disposal of Fly Ash Notification (1999):* The main objective of this notification was to conserve the topsoil, protect the environment and prevent the dumping and disposal of fly ash discharged from lignite-based power plants. The salient feature of this notification is that no person within a radius of 50 km from a coal- or lignite-based power plant shall manufacture clay bricks or tiles without mixing at least 25% of ash with soil on a weight-to-weight basis. For the thermal power plants the utilization of the fly ash would be as follows:

(a) Every coal- or lignite-based power plant shall make available ash for at least 10 years from the date of publication of the notification without any payment or any other consideration, for the purpose of manufacturing ash-based products such as cement, concrete blocks, bricks, panels, or any other material or for construction of roads, embankments, dams, dykes, or for any other construction activity.

(b) Every coal- or lignite-based thermal power plant commissioned subject to environmental clearance conditions stipulating the submission of an action plan for full utilization of fly ash shall, within a period of 9 years from the publication of this notification, phase out the dumping and disposal of fly ash on land in accordance with the plan.

Convention on Climate Change

The discussion of convention on climatic change began formally at the international level in 1988 by the establishment of Intergovernmental Panel on Climate Changes (IPCC) by United Nations Environmental Programme (UNEP) and World Meteorological Organization (WMO). It is a panel consisting of thousands of scientists from hundreds of countries established to assess the current state of knowledge about climate change. The IPCC's first assessment report played an important role in adoption of the United Nations Framework Convention on Climate Change (UNFCCC) at Rio-Earth summit in 1992. The convention agreed to the goal of stabilizing the greenhouse gas level in atmosphere, starting by reducing greenhouse gas emission to 1990 level by the year 2000 in all industrialized nations. As the treaty was lacking specific emission targets, the countries were to achieve the goals by voluntary means. The convention made it clear that the developed countries must take lead in combating the climate change and the adverse effects thereof. It was also emphased that the parties must take precautionary measures and should cooperate and promote supportive open economic systems leading to sustainable development.

Obligations on Developed Countries

1. To return to 1990 level of GHGs by the year 2000
2. To prepare a national communication with specific information on policies
3. To promote, facilitate, and finance the transfer of environmentally sound technologies (EST's) to developing countries

Obligations on Developing Countries

1. To prepare an inventory of GHGs
2. To prepare a general list of the steps to implement the convention
3. To undertake sustainable development as per the convention provisions

The convention also designed Global Environment Facility (GEF) as an interim mechanism for the financial assistance to support its implementation in developing countries. The convention also established institutional mechanism for periodic review and update of commitments, including the scheduling of regular conferences. Promoted by coalition of island nations, the third conference of parties to the UNFCCC met in Kyoto Japan in December 1997 to craft a binding agreement on reducing greenhouse gas emissions. In this protocol (Kyoto protocol) the parties agreed to reduce emission of six greenhouse gases to 5.2% below 1990 levels, to be achieved by 2012.

Convention on Biodiversity/Biological Diversity (CBD)

Although CITES provides some protection to some species of flora and fauna but it is inadequate to address broader issues pertaining to biological diversity. After several years of recognition, the Convention on Biological Diversity (CBD) was drafted and it became one of the pillars of the 1992 Earth summit in Rio-de Jenerio. The Biodiversity treaty as the convention is called as, was ratified in December 1993 and is now in force. The treaty established a Conference of Parties (COP) as the agency that will provide oversight and report on its task during its periodic meetings (Wang 2001).

Objectives of the Convention

1. Conservation of biological diversity
2. Sustainable use of the components of biodiversity
3. Fair and equitable sharing of benefits arising out of the utilization of genetic resources

Obligations on the Signatory Countries must:

- Adopt specific national biodiversity action plans and strategies

- Establish a system of protected areas and ecosystems within their respective countries
- Establish policies that provide incentives to promote sustainable use of biological resources
- Restore habitats that have been degraded
- Protect threatened species
- Respect and preserve the knowledge and practices of indigenous peoples
- Respect the ownership of genetic resources by countries and share the technologies developed from those resources
- Promote the awareness about the importance of biological diversity through media and educational programs
- Protect and conserve biodiversity as it stays within the sovereignty of the nations

United Nations Convention on the Law of Seas

There are certain areas of land, water, and air that do not fall in the national boundaries called as country's exclusive economic zones. The United Nation Convention on the Law of Seas (UNCLOS) held on December 10, 1982, established certain duties regarding the marine environments with the following as the important ones:

1. The member countries of the convention have strict obligations to protect and preserve the marine environments and take necessary measures for pollution control.
2. The member countries should regulate the activities such that they are not harmful to the seas.
3. The nations should cooperate at global and regional levels with international organizations to formulate rules, standards, guidelines, and procedures to protect marine environments.
4. Promotion of scientific research and data exchange preferences on marine environments.
5. It calls upon the nations to establish appropriate scientific criteria for the formulation of international environmental rules, standards, practices, procedures, and safeguards for prevention, reduction, and control of pollution caused to seas.

Space Treaty

The outer space treaty signed in 1967 commands the parties to pursue studies of outer space including the moon and other heavenly (celestial) bodies. It permits conducting exploration of these heavenly bodies, so as to avoid harmful contamination and adverse changes in earth's environment due to introduction of extraterrestrial matter and adopt necessary measures for this purpose.

Convention on Forests

The issue of an international forestry convention with an international binding character, raised by European Economic Commission (EEC) led by Germany and supported by the United States, was very much controversial. The EEC and the United States pursued a global forest treaty which would have been a binding treaty. Developing countries strongly opposed and challenged the binding treaty for that would have meant compromising with their sovereign rights to exploit their own resources. They argued that the developed world was trying to globalize the natural resources of the developing countries while allowing its own forests to be ruthlessly exploited on the plea that these were privately owned. India and other developing countries strongly opposed the move and took the stand that our forests could no longer be available as grounds for dumping hazardous wastes by the developed world. The convention was finally signed by various countries reflecting the first global convention on forests. It was also decided to keep the forests under regular assessment for International Corporation.

The United Nations Conference on Environment and Development (UNCED) also known as "Earth Summit" took place in June 1992, in Rio de Janerio, the capital of Brazil. It marked the 20th anniversary of the first United Nations conference on Human Environment (UNCHE) which took place between June 5–10, 1992 in Stockholm Sweden. The conference was attended by the leaders and representatives from 180 countries. Sustainable development was the primary focus of this summit. The UNCED was guided by the remarkable document of 1987, i.e., the Brundtland report and ended with the adoption of certain crucial documents—a "blueprint" intended to guide development in sustainable direction into and through the twenty-first century.

Rio Declaration (Earth Charter)

It is a document comprising the following 27 principles, intended to generate the international cooperation and global partnership for the promotion of sustainable development:

1. Human beings are at the center of sustainable development and are entitled to a healthy and productive life in harmony with nature.
2. States have the sovereign right to exploit their own resources with the responsibility to ensure the safety of their environment and of the areas beyond their national jurisdiction.
3. The right to development must be fulfilled sustainably.

4. Environmental protection shall constitute an integral part of the developmental process.
5. To adopt policies for combating poverty.
6. Environmentally vulnerable areas should be given priority in the field of environmental protection.
7. States shall cooperate to conserve, protect, and restore the health and integrity of earths ecosystems.
8. States shall take up measures for changing consumption patterns and promotion of appropriate demographic policies.
9. Measures for exchange of scientific and technological knowledge.
10. Handling of environmental issues by the participation of all citizens at the relevant levels.
11. Implementation of effective environmental legislation.
12. Adoption of a supportive and open economic system at international level.
13. National laws regarding liability and compensation of pollution and other environmental damages to victims.
14. Discouragement and prevention of the transfer of any hazardous substance or activity to other states.
15. The precautionary approaches shall be widely adopted.
16. National authorities should endeavor to promote internalization of costs and the use of economic instruments.
17. EIA as a national instrument shall be undertaken whenever required.
18. Information about natural disasters and other emergencies shall be immediately given to other states.
19. Information about the activities that may have trans-boundary environmental impacts shall be provided to the other states.
20. Women should be involved to forge a global partnership.
21. Involvement of youth to forge a global partnership.
22. Involvement of locals and their traditional practices in the environmental management and development.
23. Environment and natural resources of people under oppression, domination, and occupation shall be protected.
24. Respecting the international laws providing protection to the environment in times of armed conflicts.
25. Peace, development, and environmental protection are interdependent.
26. Resolution of environmental disputes peacefully as per the UN charter.
27. Cooperation between states and people in good faith and in spirit of partnership.

Agenda-21

It is not a legally binding document but a "work plan" or "agenda for action" or "action plan" for the twenty-first century in all areas of environment and economic growth in a sustainable way. It is a document which embraces the entire environmental and developmental agenda. Agenda-21 has four main sections.

Section 1 (Social and Economic Dimensions)
Its main focus is the sustainable development of developing countries by better domestic policies and economic systems that stresses on:

- Changing consumption pattern
- Better demographic dynamics
- Better human health
- Better human settlements

It also includes the need to integrate the environmental factors into laws, policy making, accounting and economic instruments.

Section 2 (Conservation and Management of Resources for Development)
This section promotes the understanding of integrated planning and management of various natural resources like:

- Land resources
- Mountains
- Water resources
- Agriculture
- Biological diversity
- Forests

It also focuses on the need for better information on risk assessment and management of toxic chemicals. Furthermore, it includes waste minimization, recycling, environmentally sound disposal of solid wastes, sewage, and radioactive wastes.

Section 3 (Strengthening the Role of Major Groups)
This section includes entirely the statements on the importance of following nongovernmental sector in implementing sustainable development of women, children, and youth, local authorities, business groups, trade unions, farmers, science, and technologies.

Section 4 (Means of Implementation)
This section discusses the establishment of a sustainable development commission; a new body under the Economic and social council, to coordinate the pursuit of sustainable development among international organizations and to monitor progress by the nongovernmental organizations, governmental organizations, and international organizations for the achievement of the goals of Agenda-21. It discusses the international legal instruments and mechanisms. It also discusses the issue of promoting public awareness of environmental issues through education and training, cooperation for capacity building in developing countries for implementing Agenda-21.

International Conventions

United Nations Conference on the Human Environment (Stockholm Conference)

The United Nations Conference on Human Environment (also known as the Stockholm Conference) was an international conference convened under United Nations auspices held in Stockholm, Sweden from June 5 to 16, 1972. It was the UN's first major conference on international environmental issues, and marked a turning point in the development of international environmental politics (Bradnee 1998).

Weden first suggested to United Nations Economic and Social Council (ECOSOC) in 1968, the idea of having a UN conference to focus on human interactions with the environment. ECOSOC passed resolution 1346 supporting the idea. General Assembly Resolution 2398 in 1969 decided to convene a conference in 1972 and mandated a set of reports from the UN secretary-general suggesting that the conference focus on "stimulating and providing guidelines for action by national government and international organizations" facing environmental issues (Stavins et al. 2014).

The meeting agreed upon a declaration containing 26 principles concerning the environment and development; an Action Plan with 109 recommendations, and a Resolution. Principles of the Stockholm Declaration are:

1. Human rights must be asserted, apartheid and colonialism condemned.
2. Natural resources must be safeguarded.
3. The Earth's capacity to produce renewable resources must be maintained.
4. Wildlife must be safeguarded.
5. Non-renewable resources must be shared and not exhausted.
6. Pollution must not exceed the environment's capacity to clean itself.
7. Damage to oceans due to pollution must be prevented.
8. Development is needed to improve the environment.
9. Developing countries therefore need assistance.
10. Developing countries need reasonable prices for exports to carry out environmental management.
11. Environment policy must not hamper development.
12. Developing countries need money to develop environmental safeguards.
13. Integrated development planning is needed.
14. Rational planning should resolve conflicts between environment and development.
15. Human settlements must be planned to eliminate environmental problems.
16. Governments should plan their own appropriate population policies.
17. National institutions must plan development of states' natural resources.
18. Science and technology must be used to improve the environment.
19. Environmental education is essential.
20. Environmental research must be promoted, particularly in developing countries.

21. States may exploit their resources as they wish but must not endanger others.
22. Compensation is due to states thus endangered.
23. Each nation must establish its own standards.
24. There must be cooperation on international issues.
25. International organizations should help to improve the environment.
26. Weapons of mass destruction must be eliminated.

One of the seminal issues that emerged from the conference was the recognition for poverty alleviation for protecting the environment. The Indian Prime Minister Indira Gandhi in her seminal speech in the conference brought forward the connection between ecological management and poverty alleviation (Reiner 1998).

Some argue that this conference, and more importantly the scientific conferences preceding it, had a real impact on the environmental policies of the European Community (that later became the European Union). For example, in 1973, the EU created the Environmental and Consumer Protection Directorate, and composed the first Environmental Action Program. Such increased interest and research collaboration arguably paved the way for further understanding of global warming, which has led to such agreements as the Kyoto Protocol and the Paris Agreement, and has given a foundation of modern environmentalism.

Kyoto Protocol

The Kyoto Protocol is an international agreement linked to the United Nations Framework Convention on Climate Change, which commits its Parties by setting some internationally binding emission reduction targets. Recognizing that developed countries are principally responsible for the current high levels of greenhouse gas emissions in the atmosphere as a result of more than 150 years of industrial activity, the Protocol places a heavier burden on developed nations under the principle of "common but differentiated responsibilities." The Kyoto Protocol was adopted in Kyoto, Japan, on December 11, 1997, and came into force on February 16, 2005. The detailed rules for the implementation of the Protocol were adopted at COP 7 in Marrakesh, Morocco, in 2001, and are referred to as the "Marrakesh Accords." Its first commitment period started in 2008 and ended in 2012. On December 21, 2012, the amendment was circulated by the Secretary-General of the United Nations, acting in his capacity as Depositary, to all Parties to the Kyoto Protocol in accordance with Articles 20 and 21 of the Protocol (Michael et al. 1999).

During the first commitment period, 37 industrialized countries and the European Community committed to reduce greenhouse gases emissions to an average of 5% against 1990 levels. During the second commitment period, parties committed to reduce greenhouse gas emissions by at least 18% below 1990 levels in the 8-year period from 2013 to 2020; however, the composition of parties in the second commitment period is different from the first.

Under the protocol it was attributed that the countries must meet their targets primarily through national measures. However, the Protocol also offers them an additional means to meet their targets by way of three market-based mechanisms including International Emission Trading (IET), Clean Development Mechanism (CDM), and Joint implementation (JI) to stimulate green investment and help parties to meet their emission targets in a cost-effective way.

Monitoring Emission Targets

Under the protocol, countries' actual emissions have to be monitored and precise records of the trades carried out have to be kept.

Registry system track and record transactions by parties under the mechanisms. The UN Climate Change Secretariat, based in Bonn, Germany, keeps an international transaction log to verify that transactions are consistent with the rules of the protocol.

Reporting is done by parties by submitting annual emission inventories and national reports under the protocol at regular intervals.

A compliance system ensures that parties are meeting their commitments and helps them to meet their commitments if they have problems in doing so.

Adaptation

The Kyoto Protocol, like the other conventions is also designed to assist countries in adapting to the adverse effects of climate change. It facilitates the development and deployment of technologies that can help increase resilience to the impacts of climate change.

The Adaptation Fund was established to finance adaptation projects and programs in developing countries that are parties to the Protocol. In the first commitment period, the Fund was financed mainly with a share of proceeds from CDM project activities. In Doha (2012), it was decided that for the second commitment period, international emission trading and joint implementation would also provide the adaptation fund with a 2% share of proceeds.

The Road Ahead

The Kyoto Protocol is seen as an important first step toward a truly global emission reduction regime that will stabilize GHG emissions, and can provide the architecture for the future international agreement on climate change.

In Durban, the Adhoc Working Group on the Durban Platform for enhanced action was established to develop a protocol, another legal instrument or an agreed outcome with legal force under the Convention, applicable to all Parties. The ADP was to complete its work as early as possible, but no later than 2015, in order to adopt this protocol, legal instrument or agreed outcome with legal force at the 21st session of the Conference of the Parties and for it to come into effect and be implemented from 2020.

Montreal Protocol

The Montreal Protocol on substances that deplete the ozone layer (a protocol to the Vienna Convention for the Protection of the Ozone Layer) is an international treaty designed to protect the ozone layer by phasing out the production of numerous substances that are responsible for ozone depletion. It was agreed on August 26, 1987, and come into force on August 26, 1989, followed by a first meeting in Helsinki in May 1989. Since then, it has undergone eight revisions, in 1990 (London), 1991 (Nairobi), 1992 (Copenhagen), 1993 (Bangkok), 1995 (Vienna), 1997 (Montreal), 1998 (Australia), 1999 (Beijing), and 2016 (Kigali, adopted, but not in force). As a result of the international agreement, the ozone hole in Antarctica is slowly recovering and it is projected that the ozone layer will return to 1980 levels between 2050 and 2070. Due to its widespread adoption and implementation it has been hailed as an example of exceptional international cooperation, with Kofi Annan quoted as saying that "perhaps the single most successful international agreement to date has been the Montreal Protocol." In comparison, effective burden sharing and solution proposals mitigating regional conflicts of interest have been among the success factors for the Ozone depletion challenge, where global regulation based on the Kyoto Protocol has failed to do so. In this case of the ozone depletion challenge, there was global regulation already being installed before a scientific consensus was established. Also, overall public opinion was convinced of possible imminent risks (UNEP 1999).

The two ozone treaties have been ratified by 197 parties, which include 196 states and the European Union, making them the first universally ratified treaties in United Nations' history.

These truly universal treaties have also been remarkable in the expedience of the policy making process at the global scale, where bare 14 years lapsed between a basic scientific research discovery (1973) and the international agreement signed (1985 and 1987). When comparing this very success story with attempts to establish an international policy on the Earth's climate or atomic energy, the entire process from a problem formulation to a global acceptance supported by a legal framework took less than a quarter of a single human generation life span.

Terms and Purposes

The treaty is structured around several groups of halogenated hydrocarbons that deplete stratospheric ozone. All of the ozone-depleting substances controlled by the Montreal Protocol contain either chlorine or bromine. Some ozone-depleting substances (ODSs) are not yet controlled by the Montreal Protocol, including nitrous oxide (N_2O). For each group of ODSs, the treaty provides a timetable on which the production of those substances must be shot out and eventually eliminated.

Progress of Ozone-Depleting Substances Phaseout in India

India is in the process of phasing out ODSs both in the end-use consumption sector and production sector. A total of 347 projects in the consumption sector have been approved and funded by the Multilateral Fund and of these, 270 are ozone-depleting substances (ODS) phaseout investment projects while 77 are noninvestment and support activities. A total amount of about USD 127 million has been approved by the Executive Committee of the Multilateral Fund Secretariat for phasing out 12,243 tons of ODS's. The Executive Committee of the Multilateral Fund approved a total of $ 82 million for the phased reduction and cessation of the entire CFC production in India. In this project, it has been agreed to reduce total CFC production in accordance with an agreed upon schedule. A Project Management Unit (PMU) is operational in the Ozone Cell for monitoring CFC production phaseout and implementing other support activities to aid CFC production phase out. So far, the CFC producers have achieved a reduction of 3718 MT of ODS production since calendar year 2000 and have complied with their respective production quotas. The Executive Committee of the Multilateral Fund also approved $ 2.6 million for phasing out halon production and remaining consumption of halons. The enterprises producing halons are in the process of dismantling their plants to render them unusable for halon production.

India's HCFC Phaseout Management Plan (HPMP)

Under the Montreal Protocol, the accelerated phaseout of hydrochlorofluorocarbons (HCFCs) is underway with an aim to complete phase out by 2030 of these chemicals that result in ozone depletion and aid in global warming. At present, HCFCs are used in various sectors like refrigeration and air conditioning (RAC), polyurethane foam manufacturing and cold chains sectors, etc. These sectors are directly related to urban development, agriculture through cold chain, and industrial development. India is undertaking phaseout of HCFCs through the implementation of HPMP (Jason 2000).

HCFC Phaseout Management Plan Stage-I

The 19th Meeting of the Parties (MOP) to the Montreal Protocol on substances that deplete the Ozone Layer, in September 2007 decided to accelerate the phaseout of HCFC's by 10 years with certain reduction steps both in Article 5 and non-Article 5 Parties for the early recovery of Ozone layer and to save the climate system.

The Ozone Cell, MoEF organized a consultative stakeholders meeting on June 4, 2008 to discuss India's HCFCs Phaseout Management Plan (HPMP) and the activities to be initiated to determine the base level production and consumption of HCFCs. Subsequently, India submitted the proposal for preparation of HPMP for the consideration of 56th Ex-Com held in November 2008. The same was approved by the Ex-Com with a total funding of $ 570,000 (Jorgen 1999).

The Ozone Cell, MoEF also developed a Roadmap for Phaseout of HCFCs in India describing the long term vision and action plan including the policy instruments for phaseout of production and consumption of HCFCs in India in accordance with the decision of the 19th MOP. The Road Map for the implementation of phaseout of HCFCs in India was launched in October 2009. A Sectoral Working Groups Meeting organized on September 24–25, 2009 which was attended by the stakeholders from industry, research institutions, government organizations, NGOs, and implementing agencies to develop an action plan for implementation of the Roadmap for HCFC phaseout. The Action Plan listed the major initiatives and actions along with implementing organizations/agencies and timelines.

The HPMP was prepared in close cooperation with industry associations. The Memorandum of Agreements (MOAs) were signed between the Ozone Cell, MoEF and Refrigeration and Air-conditioning Manufacturers Association (RAMA) and Indian Polyurethane Association (IPUA) for preparation of RAC manufacturing and Foam manufacturing Sectoral Strategies. RAMA and IPUA organized awareness workshops in close cooperation with Ozone Cell, MoEF, one in each major industrial hubs like Chennai, Delhi, and Mumbai, in June/July 2010. These workshops were attended by stakeholders from Foam and RAC Sectors. RAMA and IPUA made extraordinary efforts to involve Small and Medium Enterprises (SMEs) in these workshops to create awareness among SMEs. RAMA and IPUA also carried out detailed surveys involving market research consulting agencies for collection of data of number of enterprises using HCFCs, the date of establishment and annual consumption of HCFCs for the past 3 years. The information was collated and analyzed by RAMA and IPUA through their sub-sectoral committees and sectoral strategies and submitted to the Ozone Cell in March/April 2011 (Kroeze 1995).

The RAC Servicing strategy was prepared by the Servicing Sector Group of the industry under the guidance of GIZ, Government of Germany as implementing agency in close cooperation with the Ozone Cell, MoEF.

A two-day stakeholder workshop was organized on October 21–22, 2011 for finalization of sectoral strategies and overarching HPMP. A large number of stakeholders especially from industry association, defense forces, NGOs, R&D organizations, and implementing agencies participated actively in the deliberation and provided their inputs to the HPMP. The HPMP was finalized by the lead implementing agency, UNDP in association with other implementing agencies, UNEP, UNIDO and GIZ and in close cooperation with Ozone Cell, MoEF and submitted to the Multilateral Fund for Implementation of the Montreal Protocol for consideration by the 66th Executive Committee (Ex-Com). The HPMP Stage-I is aimed at phaseout of certain HCFCs to meet the 2013 freeze and 10% reduction in 2015. The HPMP Stage-I of India was approved by the 66th Ex-Com in its meeting held in April, 2012 in Montreal, Canada for the period 2012–2015 to reduce HCFC consumption to meet the 2013 and 2015 targets. The HPMP Stage-I was expected to address the conversions in foam manufacturing sector from HCFCs to non-ODS technologies in the enterprises with large consumption of HCFC-141b, system houses and activities in the Refrigeration Air-Conditioning (RAC) servicing sector.

Stage II of HCFCs Phaseout Management Plan (HPMP)

The Union Ministry of Environment, Forests and Climate Change (MoEFCC) launched the Stage II of HCFCs Phase out Management Plan (HPMP) for the period of 2017–2023.

It aimed to phase out the use of hydrochlorofluorocarbons (HCFCs), harmful ozone-depleting substances (ODS) by switching over to non-ozone depleting and low global warming potential technologies.

Under HPMP-II:

- India secured $44.1 million from Multilateral Fund for implementation of Montreal Protocol for phasing out 8190 MT of HCFC consumption between 2017 and 2023 to meet targets under the protocol for 2020.
- More than 400 enterprises, including MSMEs in foam manufacturing sector and six large air-conditioning manufacturing enterprises were supported for conversion to non-HCFC technologies from HCFCs.
- Promotion of energy efficiency, development building codes, cold chain development with non-HCFC alternatives and development of standards for new non-ODS and low GWP alternatives.
- Adequate attention to synergize the Refrigeration and Servicing (RAC) servicing sector trainings with the Skill India Mission, in order to multiply the impact of skilling and training.
- Training of nearly, 16,000 service technicians under HPMP-II to reduce the net direct CO_2-equivalent emission of about 8.5 million metric tons annually from 2023.

References

Bakshi, P. M. (1993). *The AIR (Prevention and control of pollution) act, 1981: A study* (pp. 1–79). New Delhi: Indian Law Institute.

Bradnee, W. C. (Ed.). (1998). *Global climate governance: Inter-linkages between the Kyoto protocol and other multilateral regimes.* Tokyo: United Nations University.

Iyer, V. R. K. (1984). *Environmental pollution and the law.* Indore: Vedpal Law House.

Jaiswal, P. S., & Jaswal, N. (2003). *Environmental law.* Delhi: Pioneer Publications.

Jason, A. (2000). *Keeping cool without warming the planet: Cutting HFCs, PFCs, and SF6 in Europe.* Brussels: Climate Network Europe.

Jorgen, F. (1999). HFC, PFC and SF6 emission scenarios: Recent developments in IPCC special report on emission scenarios. In *Paper presented at the joint IPCC/TEAP expert meeting on options for the limitation of emissions of HFCs and PFCs*, Petten, the Netherlands, pp. 26–28.

Juergensmeyer, J. C. (1971). *Common law remedies and protection of the environment* (Vol. 6, pp. 215–236). Atlanta, GA: Georgia State University College of Law, Faculty Publications.

Kiss, A., & Shelton, D. (2004). *International environmental law*. Ardsley: Transitional Publisher.

Kramer, L. (2006). *Environmental law* (6th ed.). London: Sweet & Maxwell.

Kroeze, C. (1995). *Fluorocarbons and SF6: Global emission inventory and options for control. No. 773001007*. Bilthoven: National Institute of Public Health and Environmental Protection.

Kumar, K. J. (1992). Environmental acts: A critical overview. In Leelakrishnan (Ed.), *Law and environment*, p. 237.

Kumari, V. K. B. (1984). Environmental pollution and common law remedies. *Cochin University Law Review, 8*, 101–114.

Leelakhishnan, P. (2002). *Environmental law in India*. New Delhi: Butterworth.

Leelakhishnan, P. (2008). *Environmental law in India*. New Delhi: Leisnexix.

Michael, G., Vrolijk, C., & Brack, D. (1999). *The Kyoto protocol. A guide and assessment*. London: RIIA/Earthscan.

National Informatics Centre, Jammu & Kashmir, (NICJK). (2000). Jammu & Kashmir Wildlife (Protection) Act, 1978 The Jammu and Kashmir Wildlife (Protection) Act, 1978 [Act. No. VIII of 1978].

Nayak, R. K. (1999). *International environmental law*. Joondalup, WA: Edith Cowin University.

Reiner, G. (1998). The strange success of the montreal protocol: Why reductionists accounts fail. *International Environmental Affairs, 10*(3), 197–220.

Rodgers, W. H., Jr. (1977). *Hand book on environmental law* (p. 100). London: Routledge.

Rodgers, W. H., Jr. (1982). Bringing People Back: Toward a Comprehensive Theory of Taking in Natural Resources Law, 10 ECOLOGY L.Q. 205, 220.

Sahasranaman, P. B. (2009). *Handbook of environmental law*. New Delhi: Oxford University Press.

Stavins, R.,& Zou, J., et al. (2014). International cooperation: Agreements and instruments. Archived 29 September 2014 at the Wayback machine chapter 13. In *Climate change 2014: Mitigation of climate change. Contribution of Working Group III to the Fifth Assessment Report of the Intergovernmental Panel on Climate Change*. Cambridge: Cambridge University Press.

UNEP. (1999). *The implications to the Montreal protocol of the inclusion of HFCS and PFCS in the Kyoto protocol (by HFC/PFC task force of the technology and economic assessment panel)*. Nairobi: UNEP.

Wang, S. (2001). Towards an international convention on forests: Building blocks versus stumbling blocks. *The International Forestry Review, 3*(4), 251–264.

Weinberg, P., & Reilly, K. A. (2013). *Understanding environmental law* (p. 458). Danvers, MA: LexisNexis or Matthew Bender & Company.

Management of Natural Resources

5

Abstract

The natural resources include any and every such material that can be transformed in a way that the material becomes extra suitable, valuable, and useful to man or it is anything obtained from nature to meet our needs and wants. Nature is a regular source of a variety of substances required to meet our basic needs in our day-to-day life. Even the prehistoric man who lived a hunter-gatherer type of life procured each and everything for his survival from nature which has been kind enough to him since that time. Although natural resources like land, air, water, forests, wildlife, minerals, metals, energy, and various other resources existed over the earth's surface even during prehistoric times, man had neither the tools nor the technology to use them. But with time he learnt the process of cultivation of land, growth of crops, operation of wind and water mills and many other techniques and technologies that are required to harness these resources to the fullest of their potential. However, the expanding human population alongside of its technological development and scientific progress started utilizing the natural resources at a much faster rate and a larger scale. This continuous increase in human population caused an expanding demand for these resources thus creating a situation when the non-renewable natural resources may come to an end after sometime. In order to have maximum production we have actually started taking loans from the resources meant for future that cannot be paid back. As a result we would be using all those resources which are in fact the property of the future generations which is a matter of great concern. So there must be some sort of balance between the population growth and their utilization by adopting appropriate management strategies of these natural resources for making the planet sustainable.

Keywords

Natural resources · Land resources · Energy · Wildlife · Forests

Natural resource management, that specifically focuses on the technical and scientific understanding of the resources like soil, water, land, plants, animals, their ecology and life supporting capacity, deals with the management of the way of interaction between people and resources.

It brings together water management, land-use planning, biodiversity conservation, and the future sustainability of industries like mining, tourism, agriculture, forestry, and fisheries. It recognizes that people and their livelihoods rely on the health and productivity of the landscapes, and their actions as stewards of the land play a critical role in maintaining their health and productivity (Williams 2011).

Soil Conservation

Soil is the most fundamental resource to fulfill the basic requirements of food, fiber, and shelter of human race. It is the basis of all terrestrial life and provides a wide range of ecosystem services such as pollutant filtration, purification, retention of sediments and chemicals, buffering and transformation. Conservation of soil is very important for sustainability of agriculture and environment as it is under immense pressure due to the ever increasing human population. Soil is being deteriorated by many anthropogenic and natural factors like soil erosion and soil pollution. Soil loss due to erosion has a greater consequence as it leads to loss of productivity. It occurs throughout the world but it is a very common and serious problem in dry areas (Govers et al. 2017). Soil erosion disturbs agricultural, environmental, and ecological functions of the soil. It results in the depletion of soil fertility, reduction of moisture storage capacity, and consequently reduction in crop productivity. In addition to the loss of fertility and crop yields, soil erosion also increases environmental pollution by increasing the sediment load in streams and rivers, thereby disturbing the aquatic life as well.

Nowadays, accelerated erosion due to misuse of resources like land, water, and soil is seen as the most pressing problem before man. Further, vast tracts of fertile land are rendered useless on account of industrialization and development. It results in deterioration of the soil and a significant loss of soil productivity, thus leading to desertification under sever conditions. In order to protect the soil from erosion and to maintain the productivity of soil we can use different biological and mechanical methods of conservation (Chapin et al. 2011).

Methods of Soil Conservation

The practical methods of soil conservation can be broadly divided into two types.

Biological Methods

Biological methods of soil conservation involve the protective action of the vegetation cover for the conservation of soil. The dense vegetation cover helps in the:

- Prevention of splash erosions
- Reduction of surface water runoff velocity and increase of infiltration rate due to increased surface roughness
- Facilitation of soil particle accumulation
- Stabilization of soil aggregates through the action of roots and organic matter

The major biological methods useful for the checking the erosion of soil are as follows:

I. Agronomic methods
II. Agrostological methods
III. Dry farming practices

I. Agronomic Methods

Agronomic practices/methods are the practices used by farmers to improve the quality of soil, increase the productivity of cultivated land, enhance the water usage, and improve the environment. They are a vital part of the farming systems which contribute to the conservation of soil and include the following principal techniques of soil conservation:

1. Contour farming
2. Conservational tillage
3. Crop rotation
4. Mulching
5. Strip cropping

1. Contour Farming

Also known as contour bunding or contour plowing, it is a farming practice in which all the cultivation operations are done on the contour lines, which create the water breaks and reduce the formation of gullies and rills during the lines of heavy runoff (Panos et al. 2015). It provides for the conservation of rainwater and reduction of soil loss by stopping the process of erosion. Its purpose is to place rows and tillage lines across the normal flow on the surface which usually gains velocity on the steep slopes. These ridges and furrows thus created reduce the runoff velocity and give more time for the water to infiltrate into the soil (Govers et al. 2017). It is of an increasingly high value in arid and semiarid regions where moisture is usually a limiting factor in providing a satisfactory crop cover against erosion.

2. Conservational Tillage

Conservational tillage, a traditional farming technique involving the planting, growing, and harvesting of crops with limited disturbance of soil, is a method of cultivation in which the previous year's crop residues are left on the field before and after the plantation of the next crop, to reduce the runoff and subsequent soil erosion along with other benefits like carbon sequestration. This practice conserves the water, soil, and energy resources through the reduction of tillage intensity and retention of crop residues. Tillage, that is the mechanical manipulation of soil by different implements like harrowing, plowing, and cultivation makes soil friable and loose to help in the retention of water (Panos et al. 2015). This type of soil tillage is characterized by tillage depth and the percentage of surface area disturbed (Arnalds 2005). Conservation tillage methods include zero-till, ridge-till, mulch-till, and strip-till. Zero-tillage is the extreme form of conservation tillage resulting in minimal disturbance to the soil surface.

3. Crop Rotation

Crop rotation, the planned sequencing of crops is an important method for checking the soil erosion and maintaining the productivity of soil. It involves a rotation of densely planted small grain crops one after the other in a set cycle on the same field over a defined period to check the soil erosion and increase the profits of productivity without impairing soil fertility. Selection of crops for rotation should be made taking into consideration the economic condition, climate, soil texture, slopes, soil types and nature of erosion etc. Deep-rooted crops should be rotated with shallow-rooted crops (Panos et al. 2015).

4. Mulching

Any organic or mineral matter such as straw, paddy husk, saw dust, ground nut shells, leaves, crop residues, paper, and loose soil applied to soil for conservation purposes is known as mulch. It is applied to the soil to reduce the impact of rains, excessive evaporation, surface crusting and hence the conservation of soil. Mulch farming that involves the leaving over of mulch on the ground surface to conserve the soil moisture, improve the soil health and fertility and reduce the weed growth can be used in higher periods/regions for decreasing water and soil loss and in lower rainfall periods/regions for enhancing the soil moisture capacity. The natural sources of mulch are the agricultural by-products for instance, the stubble, straw, manures, wood chips, and corn cobs (Loch 2004).

Different advantages of mulching are:

- Protection of soil from erosion
- Retention of soil moisture
- Compaction reduction due to heavy rains
- Reduction of the need of frequent watering
- Maintenance of a more even soil temperature
- Prevention of weed growth

- Improvement in the soil conditions by acting as a resource of organic matter and keeping the soil loose

5. Strip Cropping

It involves the cultivation of crops in alternate strips parallel to one another where some strips are allowed to lie fallow while the others are brought under the crop cultivation. As different crops are harvested at different intervals, it ensures that at no time of the year the entire area is left bare or exposed. The tall growing crops act as wind breaks and the strips which are often parallel to the contours help in increasing water absorption by the soil by slowing down runoff.

Purposes

- Reduction of soil erosion
- Reduction of the transportation of sediments and waterborne contaminants
- Protection of crop damage from windborne soil particles
- Improvement of water quality

Following are a few types of strip cropping.

(a) Contour Strip Cropping

It is the production of ordinary farm crops in long, relatively narrow strips of variable width on which dense erosion control crops alternate with erosion permitting crops. Practiced on the contours of the hill slopes, it slows down the water runoff during rainfall and minimizes the erosion of soil. While it is desirable to follow the true contour, but because of variation in the slope of land it is not always possible to have parallel strips with all the strips exactly at the same level (Pretty and Shah 1994).

(b) Field Strip Cropping

Places where soil is absorbent and it is impracticable to follow the true contours, a modified form of contour strip cropping known as field strip cropping is employed. It is a specialized strip cropping where crops are planted in parallel beds across the slope but do not follow the contour lines. Here in this practice beds of grasses or other close growing species are alternatively grown with the beds of cultivated crops. Field strip cropping, unless applied to very regular slopes, is a poor substitute for contour strip cropping, but is usually a step in the right direction (Ellis-Jones and Sims 1995).

(c) Wind Strip Cropping

The cropping practice that involves the production of tall and long growing crops, like maize, in alternate arrangement in straight and long but relatively parallel and narrow strips laid out right crosswise of the direction of prevailing winds regardless of the land contour is known as wind strip cropping. It is very effective in retarding the wind erosion, although it plays a limited role in water conservation. For this reason, careful weightage should be given to each method based on the advantages as for example the increased vegetative growth resulting from the moisture saved with contour strips may make them more desirable than strips designed to resist only wind.

(d) Temporary or Permanent Buffer Strip Cropping

An area of land maintained under permanent vegetation cover to help in controlling soil, air and water quality along with other environmental problems especially the land problems in this particular case is known as a buffer strip. In temporary or permanent buffer strip cropping the strips are established to take care of certain critical conditions like steep or highly eroded slopes in fields under contour strip cropping. Then buffer strips trap the sediments and enhance the filtration of pesticides and nutrients by slowing down the surface water runoff that could otherwise enter the local surface water (Loch 2004). Further the buffer strips by way of the root system of the planted vegetation help in reducing the wind erosion and stabilizing the stream banks by providing substantial protection against landslides and erosion.

Buffer strips can have several different configurations of vegetation found on them varying from simple grasses to combinations of grasses, trees, and shrubs. Areas with diverse vegetation provide more protection from nutrient and pesticide flow and at the same time provide better biodiversity among plants and animals.

II. Agrostological Methods

Cropping of grasses in a heavily eroded area of land is known as an agrostological measure. In agrostological practices, grasses are cultivated in rotation with the regular crops to increase the level of nutrients in soil, protect the soil, and produce fodder for cattle. For heavily eroded soils it is recommended to grow grasses for many years to let the soils repair naturally.

Some important agrostological practices used to check the process of erosion and to conserve the soil are as follows.

(1) Lay Farming

It is the practice of growing grasses in rotation with the agricultural crops. It helps in the improvement of soil fertility and in the binding of soil particles, together thus preventing the process of soil erosion. This practice is recommended for Nilgiris and similar places which are subjected to very severe soil erosion.

(2) Retiring the Land

Land retirement is a practice of taking the agricultural land out of production to provide for the recovery of the land for the increased productivity. It is an operation that can be used to address the water shortage problems in agricultural lands.

(3) Afforestation and Reforestation

Afforestation means growing forests at places where there were no forests before while reforestation means replanting of forests at places where the previously present forests have been destroyed by uncontrolled forest fires, excessive felling and lopping. Plantation of trees by afforestation or reforestation creates wind breaks and shelter belts in areas prone to wind erosion and help in reducing the same by conserving the soil. The trees or other plants, planted at right angles to the prevailing winds protect the bare soil from full force of wind by reducing the velocity of wind and hence decrease the erosion of soil. Wind break and shelter belt type of plantations are being done in some regions of Uttar Pradesh where the desert is encroaching upon the forested or otherwise planted land.

III. Dry Farming Practices

Dry farming, also known as dryland farming, is the practice of crop cultivation in areas with a limited rainfall and irrigation (<50 cm of annual rainfall). It depends upon the effective selection of crops, growing methods, and storage of the limited moisture to facilitate the efficient use of the limited level of moisture present in the soil. It is practiced in areas where the annual potential evaporation far exceeds the potential annual precipitation. Worldwide the dryland farming areas are characterized by a deficit between precipitation and evaporation which differs in the size of deficit and time of its occurrence. Dryland farming can be successfully employed in an area on the following few grounds:

1. Reduction in evaporation from soil surface
2. Retention of precipitation on land
3. Utilization of drought-tolerant crops

Mechanical Methods

The methods of soil conservation which make use of some engineering techniques or structures are known as mechanical methods. These methods supplement the biological methods of conservation when they are unable to do it properly in isolation and are aimed at the following objectives:

1. Reduction of the velocity of runoff water and its subsequent retention for a longer period so as to allow maximum absorption of water in the soil
2. Division of a longer slope into several small parts so as to reduce the velocity of runoff water to the minimum
3. Protection against erosion of soil by wind and water

The different mechanical methods used for the conservation of soil are as follows.

(1) Basin Leaching

Basin leaching is a mechanical method that involves the constructions of a series of small basins along the contour by means of an implement called basin blister which collect and retain rain water for a longer time and hence catch and stabilize downwardly moving soils of the slopes.

(2) Pan Breaking

In areas, where the soil becomes impervious to water and less productive due to formation of a hard sheet of clay below the surface, the productivity and water permeability can be enhanced by breaking the hard clay pans on the contour at a distance of some 5 feet. It improves the drainage and percolation of rain water and hence saves the soil from erosion and residual runoff.

(3) Subsoiling

It is the deep breaking of the subsoil by means of an implement called subsoiler which promotes the rain water absorption in the soil and makes the soil more loose and fit for the luxuriant growth of vegetation.

(4) Contour Terracing

Sometimes drainage channels or properly spaced ridges or soil mounds called as terraces, are formed along the contour (at right angles to the slope) to retain water in the soil and check the soil erosion. Those terraces are leveled areas constructed at right angles to the slope to reduce soil erosion.

(5) Contour Trenching

Contour trenching is the creation of a series of deep pits ($2' \times 1'$) or trenches across the slope at convenient distances with the soil excavated from the pits/trenches deposited along the lower edge in the form of a bund on which tree seeds are sown.

(6) Terrace Outlet

In order to reduce soil erosion and to remove excess rain water safely from the contour terraces, pipe outlets or channels, thickly covered by grasses are used.

(7) Gully and Ravine Control

Gully formation, which can degrade the quality of land by detaching the soil particles from their original places and by damaging the surface quality of soil can be check by the following methods:

(a) Creation of perimeter bunds around gullies to check the water flow through them
(b) Growing of suitable soil binding vegetation on the gullies to check the erosion
(c) Creation of diversion trenches around gullies

(8) Stream Bank Protection

Protection of stream banks with a highly vertical drop can be checked by growing vegetation on the slopes, by making the drop sloppy and by constructing stone or concrete pitches.

Wind Erosion Control

Wind erosion is controlled by reducing the force of wind erosion on erodible soil particles or by creating wind erosion-resistant aggregates or soil surfaces. The aim is to reduce the width of the area, maintain vegetation residues on the earth's surface, use stable soil aggregates, roughen the surface of land and level the land. Following few measures can be used for the control of wind erosion:

Management of Soil Water Wind erosion can be controlled by managing the soil water, using the practices of conservation tillage and terracing which subsequently helps in the reduction of evapotranspiration.

Alteration of field Length The length of eroding field can be altered by strip cropping or by installing wind breaks perpendicular to the direction of the prevailing wind to prevent the occurrence of wind erosion.

Management of Vegetation Alternating the less-wind resistant crops like cotton with more-wind resistant crops like sorghum along with the addition of crop residues to the fields can be helpful in wind erosion control.

Management of Land

Land management is the process of managing the use and development of land resources in both rural and urban settings.

Land-use planning is the general term used for a branch of urban planning encompassing various disciplines which seek to regulate land use in an efficient and ethical way, thus preventing land-use conflicts. Governments use land-use planning to manage the development of land within their jurisdictions and in doing so, the government plans for the needs of the community while safeguarding natural resources. It is the systematic assessment of land and water potential, alternatives for land use, and economic and social conditions in order to select and adopt the best land-use options. As a comprehensive plan, a land-use plan provides a vision for the future possibilities of development in neighborhoods, districts, cities, or any defined planning area.

It is an essential tool for pollution prevention and control. It refers to the different socioeconomic activities occurring in a particular area, the human behavior patterns they create, and their effects on the environment. Therefore, by appropriately defining land uses, establishing where and how they occur, as well as effectively controlling their performance and interrelation, governments can actively participate in preventing and controlling pollution (Rapport et al. 1998).

Land-Use Planning

Land-use planning refers to the process by which a society, through its institutions, decides where, within its territory, different socioeconomic activities such as housing, agriculture, industry, commerce, and recreation should take place. It includes protecting well-defined areas from development due to environmental, historical, cultural, or other similar reasons, and establishing provisions that control the nature of developmental activities. These controls determine features such as plot areas, their land consumption or surface ratio, their intensity or floor area ratio, their density or units of that activity or people per hectare, the technical standards of the infrastructure and buildings that serve them, and related parking allowances (Sayer and Maginnis 2005). In relation to pollution prevention, land-use provisions should include, the levels of gas emissions, light radiations, noise, water, solid waste discharges, and on-site or pre-disposal treatment of pollutants. All of these provisions should be included in the jurisdiction's land-use or zoning code as this code becomes the legal guide for developers, landowners, citizens, and authorities. A good system of protected areas, together with strong land-use provisions, always results in a less-polluted jurisdiction.

Planning Process

Following are a few planning practices that can help in attaining the positive environmental effects:

1. Define, make, and effectively protect "no-go" areas in recognition of their high environmental, historical, or cultural values, for their biodiversity.
2. Protection of "no-go" areas should be combined with measures that allow transfer of development rights from these areas into ones where development is acceptable.
3. Planning of industrial zones, by appropriately defining their location, design, infrastructure, regulation, and the buffers separating them from residential and other activity zones. In addition to pollution control measures this should be combined with fiscal and other incentives for remediation and resettlement on contaminated sites.
4. Establishment of land uses, densities, and intensity of development that result in increased usage of public transit, decreased usage of private vehicles, and reduced consumption of energy at the household level to diminish emissions, levels of air pollution, and energy use, and increase the effectiveness of existing pollution-control mechanisms and practices.
5. Integration of the rural and urban realms into the framework of land-use planning, so as to allows an integrated approach to the conceptualization and management of growth and land use, as opposed to a competitive approach in which urbanization is carried out at the expense of rural settings.

6. Careful establishment and modification of urban–rural boundaries, by closely linking them to available capacity for providing water and wastewater treatment, and to the area's economic linkages and commuting patterns.
7. Development of all types of urban agriculture within greenbelts, boundary territories and inside urbanized areas. Such agriculture would include green roofs, neighborhood and community farms, or large farm operations, and implementing creative incentives to make them economically feasible.
8. Establish authorized levels of gas emissions, air pollution, noise pollution, sun radiations, energy consumption, solid and waste water discharges, and similar measures for the different land uses and constructions that house them. Fining or penalizing operations that exceed these requirements.
9. On-site or pre-disposal treatment of pollutants, and granting incentives for additional, positive contributions to the environment.
10. Mandate the use of green building standards, techniques and materials, like the ones established by ASTM International (formerly the American Society of Testing and Materials).

Interaction with Other Tools and Possible Substitutes

Policy makers should define the land-use plan and its institutional setting in close relation to the tools defined in the World Bank Guidance Note series on tools for effective pollution management. The land-use plan should also be defined in relation to the following tools pertaining to governments:

- *Setting priorities*, a process that consists of determining and stating a national-level framework for integrated land management, as well as determining and stating the national government's perspectives regarding key land uses and decisions.
- *Environmental Assessment*, to ensure that the land-use plan requires developers to conduct these assessments and incorporate those conclusions into their projects prior to planning and construction approvals.
- *Strategic Environmental Assessment*, so the land-use plan is subject to this type of assessment, and the land-use plan incorporates all policies and measures of that assessment.
- *Industrial estates*, to guarantee that the land-use plan incorporates the practices, procedures, and regulations associated with the establishment and operation of the industrial estates.
- *Environmental Regulation and Standards*, Monitoring, inspection, compliance and enforcement, in that the plan and zoning code include all relevant standards.
- *Market-based instruments and taxation policies*, to ensure that the planning authorities incorporate and effectively apply the numerous land management instruments mentioned in this guidance note in order to formalize policies and procedures, and bring about a more equitable land market.

Advantages of Land-Use Planning

- Regulation in the level of emissions from transportation systems, less average commuting time, cultural flowering in new public spaces, less crime, and most importantly, a much greater capacity to be informed about one's city and a greater capacity to broadly communicate the advantages of such regulation.
- Correct land-use planning mechanisms and principles yields developed areas with densities sufficient to support mass transit, and increase the number of public transit trips while reducing their length.
- Reduced discharges of waste water and solid wastes, increased greenery in urbanized areas, and reduced emission of greenhouse gases.
- Maintenance of good relation between human health and land use as numerous studies have demonstrated the relations between transportation, land use, and public health.

Limitations of Land-Use Planning

- It can give rise to some complex phenomena that can exacerbate the pollution levels typically seen in developing countries.
- Skewed land-use planning can give rise to a number of social and environmental issues.
- Traffic and congestion patterns, frequently the result of a population commuting long distances on unfinished or unregulated mass transit systems, greatly affect air quality in urban space, and thus quality of life in urban space.

Forest Conservation

Practice of planting and maintaining the forested area for the benefits and sustainability of future generation is known as forest conservation. It involves the upkeep of the natural resources within a forest that are beneficial to both humans and the environment, as forests are vital for human life. They provide a diverse range of resources, store carbon to act as a carbon sink and produce oxygen which is vital for existence of life on the earth (Alaric and Patrick 2014). Rightly called as the earth's lungs, they help to regulate planetary climate, purify water, hydrological cycle, provide wild life habitat, absorb toxic gases, reduce global warming and noise, conserve soil, reduce pollution, mitigate natural hazards such as floods, landslides and so on. But nowadays, the forest cover is depleting rapidly due to many reasons such as an expansion of agriculture, timber plantation, other land uses like pulp and paper plantations, construction of roads, urbanization and industries. They constitute the biggest and severe threat to the forests, causing serious environmental damage.

Following are a few measures that need be adopted for the conservation of forests:

1. Matching of tree felling by tree planting programs.
2. Afforestation along river banks, highways, play grounds, parks and in areas unfit for agriculture by organizing some special programs of tree plantation like Van Mahotsav held every year in India.
3. Economization of timber and fuel wood by minimizing the wastages.
4. Preferential use of alternative resources of energy such as natural gas, biogas, etc. in place of fuel wood.
5. Protection of forest fires by the use of Modern firefighting equipments.
6. Pests and diseases of forest trees should be controlled by fumigation and aerial spray of fungicides and through other biological methods.
7. Discouragement of cattle grazing in forest areas.
8. Adoption of modern and updated forest management strategies like proper use of fertilizers, irrigation, bacterial and mycorrhizal inoculation, control of weeds, breeding of elite trees and tissue culture techniques.

Methods of Forest Conservation

Social Forestry
Social Forestry is a concept, program, and mission that aims at ensuring ecological, economic and social security to the people, particularly to the rural masses by involving the beneficiaries right from the planning stage to the harvesting stage. The objective of raising fast-growing trees is to meet the basic needs of the poor and landless people for fuel wood, fodder, small timber, green manure, fruits and other raw materials for self-consumption and cottage industries, alleviating poverty and contributing to the development of villages. It involves plantations on degraded forests close to habitations, government, private and village panchayat lands, canal banks, road sides, on the sides of railway lines and lands not suitable for agriculture (Kumar 2015).

The word Social forestry was coined by Westoby and used in the Ninth Commonwealth Forestry Congress in 1968. According to Prasad (1985), "it is forestry outside the conventional forests which primarily aims at providing continuous flow of goods and services for the benefit of people" implying that the production of forest goods for the needs of the local people is social forestry. Social forestry is social in the sense that it is socially configured i.e., it is adaptable, dynamic, and responsive to the social environment. Social forestry projects take varying forms depending on the particular environment (economic, political, ecological, and cultural) and remain flexible because of the social creativity of the participating interest groups. It is simply the forest practice which is of the people, by the people, and for the people (Pawar and Rothkar 2015). The social forestry practices include the raising of wind breaks on dry farm lands, planting of shelterbelts, planting along roadsides, village common lands, wastelands, railway lines, canal/stream banks, planting small wood lots in farm lands, reclamation of highly eroded and degraded soils and afforestation of command areas of irrigation projects.

Objectives of Social Forestry

Social forestry program started worldwide and adopted by the commission in 1976, has made some considerable differences in the overall forest cover in a very short time while benefiting both the rural and urban communities (Kumar 2015), based on the economic needs of the community aimed at improving the living conditions of rural and urban communities with the following objectives:

1. Fulfilling the basic requirements such as fuel, fodder, small timber, supplementary food and income from surplus forest products to the rural area and replacement of cow dung
2. Providing employment opportunities and increasing the family income considerable for alleviating poverty
3. Developing cottage industries in rural areas
4. Organizing the rural committees in their struggle for socioeconomic development and integration of economic gains in the distribution of their benefits to the rural society
5. Indoctrination of the values of village level self-sufficiency and self-management in the production as well as distribution of forest products with social justice
6. Playing a vital role in the reclamation of degraded lands, improvement of agricultural production, conservation of soil moisture, and prevention of environmental deterioration
7. Increasing the natural beauty of the landscape, creating recreational forests for the benefit of rural and urban committees
8. Protecting the agricultural fields against speedy winds and natural calamities
9. Solving the food problem of the rural area to a great extent by providing certain edible fruits like mango, cashew, palms, and coconut with high nutritional value
10. Utilizing the available land according to its carrying capacity

Afforestation

Afforestation is the establishment of forests on lands that have been without forests for some period of time, such as previously forested lands that were converted to range, and the establishment of forests on lands that have not been forested in the past. It is the establishment of a forest or stand of trees in an area where there was no previous tree cover. By planting trees and creating forests, many of the commercial needs of human beings are fulfilled, while not destroying what is left of the planet. Many governmental and nongovernmental organizations directly engage in programs of afforestation to create forests, increase carbon sequestration and carbon capture, and help to anthropogenically improve biodiversity (Ros-Tonen et al. 2005). It is, therefore, a practice that has been propagated by governmental and nongovernmental agencies of many countries as a way to stop overexploitation of nature and its resources especially the forest resources.

Agroforestry Agroforestry is the concept of planting trees along with the agricultural crops in croplands with the environmental benefits exceedingly vast, as planting trees is always beneficial whether it takes place in a barren land or alongside agricultural crops (Ros-Tonen et al. 2005).

Regulated and Planned Cutting of Trees

No doubt, the trees are considered as the perennial resources, but their large-scale exploitation makes their revival impossible, as one of the estimates suggests that in the world some 1600 million m^3 of wood were used for various purposes. So, for the proper management of this perennial resource the following strategies should be adopted:

1. *Clear cutting:* It is the selective cutting of the trees of same age group in an areas and its subsequent replantation. The clear cutting method is useful for those areas where the same types of trees are available over a large area.
2. *Selective cutting:* Followed on rotational basis, it is the selective cutting of mature trees only.
3. *Shelter wood cutting:* Shelter wood cutting is a process of cutting useless trees followed by medium and best quality timber trees. The time gap between these cuttings is helpful in regrowth of trees. In regulated cutting only one-tenth of the forest area is selected for use and rotational system is always followed for their protection.

The forest can be managed in such a way that a timber crop may be harvested indefinitely year after year without being depleted. This technique is called "sustained yield" method and is adopted by many countries around the world.

Control of Forest Fires

Forest fires, which start either by the natural process like lighting and frictional fires between trees due to speedy winds or by the intentional or unintentional acts of humans cause a huge loss to the forests and once it starts it becomes very difficult to control. Worldwide the forest fire are very common as according to an estimate, during the period from 1940 to 1950, in the United States alone, fires consumed an average of 21.5 million acres of timber yearly and as many as 1,175,664 cases of forest fires occurred during 1955–1964 in most cases they were induced by man in order to save forests from fire it is necessary to adopt latest techniques of firefighting. Some of the fire suppression techniques are:

(i) Development of wide fire lanes around the periphery of the fires
(ii) Spray of water or some fire retardant chemicals from back tanks and if possible by helicopters

Wildlife Management

The science of managing wildlife and its habitat, including people is known as wildlife management. When we think of wildlife management, we generally think of larger terrestrial and aquatic vertebrates such as birds and mammals but it actually refers to the management of any organism living outside the direct control of human beings including the uncultivated plants and undomesticated animals. Moreover, many people think of wildlife management only as a way of preserving or saving wildlife but true wildlife management encompasses these factors as well as many others. Wildlife management for conventional purposes may be described as indirectly influencing wildlife populations by altering food supply, habitat, density of competing populations, or occurrence of disease. It is more of an art than a science and is an integration of many disciplines of science that are applied together in an attempt to achieve a foreseeable outcome (Groom et al. 2006). It can be applied in several different forms which include direct or indirect population, manipulation and preservation which refers to the management techniques that basically allow natural processes to take their course without human intervention. Direct population manipulation implies altering populations by trapping, hunting, stocking, or shooting while the indirect population manipulation, the most widely, used method, influences populations by altering key components of a population's habitat, food, and water.

The concept of wildlife conservation which has been around since ancient times now has evolved into a science, but its goal remains essentially the same: to ensure the wise use and management of the renewable resources. For the maintenance of this renewable living resource we need the sustainable wildlife management (SWM) which is the sound management of wildlife species to sustain their populations and habitat over time, taking into account the socioeconomic needs of human populations. This requires that all land users within the wildlife habitat are aware of and consider the effects of their activities on the wildlife resources and habitat, and on other user groups (Groom et al. 2006). In view of its ecological and socioeconomic value, wildlife is an important renewable natural resource, with significance for areas such as land-use planning, rural development, tourism, food supply, cultural heritage, and scientific research. If sustainably managed, wildlife can provide continuous nutrition and income. It can considerably contribute to the alleviation of poverty as well as to safeguarding human and environmental health.

Methods of Wildlife Management

1. *Habitat management*: Habitat management is a practice that seeks to protect, conserve, and restore habitat areas for wild animals and plants, especially conservation reliant species, and prevent their extinction, fragmentation, or reduction in range. The purpose of habitat management is to improve existing habitats to benefit the wildlife. We can often increase the amount of wildlife in an area, improve their quality of health, and encourage them to use areas that they currently are not using just by manipulating the habitat. However, there are still

certain limitations to what we can do. With the expanding human populations and the growing use of resources, it is often difficult to balance our need for resources and the increasing need to conserve existing habitats and biodiversity. One of the hardest parts of habitat management is being able to establish clear goals that are feasible and compatible with surrounding land uses and future plans for the land (Hunter 1996). Habitat management can be accomplished using a variety of approaches, including preservation, conservation, or mitigation. For example, if one intends to preserve an existing tract of land as a wildlife sanctuary, it would be best if that tract can be connected to other existing tracts of land rather than creating an isolated habitat.

2. *Balancing Act*: It is the act of balancing the habitats in order to support wildlife as for example when we remove a certain population of plants and animals from a community and the community may not survive. This typically happens when urban development pushes into wildlife areas.

3. *Carrying Capacity*: Resources in any given habitat can support only a certain specific number of organisms and this capacity of the resources is known as carrying capacity. Carrying capacity of a certain tract of land can vary from year to year and can be changed by both nature and humans.

4. *Limiting Factors*: Factors like diseases and starvation, predators and hunting, accidents, pollution and old age can limit the potential production of wildlife. So for the proper management of the same we need to manage these factors sustainably so that they do not limit the potential growth of wildlife.

Hunter's Role in Wildlife Conservation Since wildlife is a renewable resource, hunters can help maintain wildlife populations at a healthy balance as it is an effective wildlife management tool because the hunters play an important role by providing information from the field to the wildlife managers. Funding from hunting licenses has helped many game and non-game species recover. In addition to participating in the harvest of surplus animals, hunters help to sustain the game populations by participating in the surveys, providing the samples from harvested animals and funding for wildlife management through license fees.

Wildlife Management Practices

Monitoring Wildlife Populations Wildlife managers continuously monitor birth and death rate of various species and condition of their habitat. This provides data needed to set hunting regulations and determine if other wildlife management practices are needed to conserve wildlife species (Soule and Wilcox 1980).

Habitat Improvement As succession occurs, change in habitat affects type and number of wild plants and animals that a habitat can support. Wildlife managers may cut down or burn forested areas to promote new growth and slow down the process of succession. This practice enables them to increase the production of certain wildlife species.

Hunting Regulations Hunting regulations, including setting daily and seasonal time limits and legal method, protects habitats and preserve animal populations.

Beneficial Habitat Management Practices Beneficial habitat management practices like food plots and planting, controlled burning, brush pile creation, timber cutting, ditching, diking, nuisance plant and animal control, and water holding needs to be adopted for the better management of the wildlife.

Wildlife Management Tools

- *Laws:* Flexible wildlife related laws, based on biological facts, must be used in combination with other management tools for the management of wildlife. These game laws are necessary for the safety of people, protection of the game, and their fair sharing with the future generations.
- *Stocking*: It is the release of wildlife species in areas that have suitable habitats but no animal population.
- *Hunting and Trapping*: It is a valuable tool for maintaining wildlife populations at or below carrying capacity for the habitat because by regulating hunting only excess animals in a population are removed.
- *Public Education*: It helps in the public understanding of wildlife management necessary for different management programs because the more people know and understand about the wildlife, the more likely they support the management programs.

Management of Landslides

Landslides, the widespread and unpredictable phenomena, are the major hydrogeological and anthropogenic hazards that affect not only the mountainous areas but also the river terrains, mining areas, offshore, and coastal areas.

Landslide hazard management in India had till now been confined to ad hoc solutions of site specific problems and the implementation of immediate remedial measures including debris removal, and dumping of this debris either down slope or into a river.

Landslide hazard management involves measures taken to avoid or mitigate the risk posed by them and the most important role in this process is played by the local government machinery. The government machinery initiates steps to warn the communities about the risk, as and when the information is received about the probability of landslide occurrence within its jurisdiction, and tries to convince landowners/dwellers to shift to safer places. Moreover, further development is avoided in such high risk zones (Singh 2010.

Mitigation strategies might not be possible in every landslide hazard prone area because of their high cost and the indifferent attitude of the public. Efforts to reduce their risks are also made by road construction and maintenance agencies by implementing required treatment measures.

There is, however, a need to preempt a disaster by making adequate information available in advance before it strikes, something that is emphasized in the disaster management guidelines which are to be used by all states, especially those affected by multi-hazards.

Rescue Measures

Rescue Operation

As a first hand responsibility, the local people should inform SDMA, police, local government body, hospitals, or media for the quick rescue operation and in case of large-scale hazards the government should take help of Para military combat force (Clerici et al. 2002).

First Aid and Hospitalization of Injured Peoples

Seriously injured landslides victims should be provided with a quick first aid and in case of any emergency, they should be admitted in the nearby hospitals to provide some specialized treatment.

Removal of Debris

Landslide debris blocks all the roads and thus brings the transport system to a halt. So, highway/roadway authority should always be prepared to remove the debris materials from the roadways as early as possible to avoid any inconvenience caused to the tourists, travelers, patients, or other commuters.

Structural Measures

Slope Stabilization

Increase in the stability of the vulnerable slopes either by expanding the volume at the toe or by reducing the volume at the head.

Drainage Pattern

(a) The streams and temporary water resources at the head of the slides should be diverted from the slide area.

(b) Periodic maintenance of drainage in pre-monsoon and monsoon periods should be designated as an essential service.

Afforestation

Afforestation programs should be lunched to minimize the breakdown situation of environmental equilibrium and the grass cover must not be disturbed unnecessarily in the hilly areas.

Construction of Human Settlements

Due to population pressures and increasing influence of tourism industry the vulnerability of landslides increases day by day throughout the world. Many heritage sites in the hilly areas are now subjected to landslide hazards. So, the agencies and institutions involved in the maintenance of these heritage sites in different countries should take due care to save them as for example in India the National Trust for Archaeological and Cultural Heritage (INTACH) and State Archaeological Departments hold follow scientific measures to save these heritage sites. Planners, architects, and engineers should design human settlements in consideration of ecological fragility of the hilly areas (NDMG 2009).

Rehabilitation of Settlements

Landslides destroy houses, kill people, block roads and bring an untold suffer to the victims. So it seems the prime role of the local governments to responsibly rehabilitate the people at an earliest.

The wide expansion of crop plantation and urbanization over the forest cover should be stopped.

Reduction in the channel gradient by building check dams with a cover of shrubby vegetation.

Unscientific cultivation and mining should be stopped.

Management of Earthquakes

An earthquake is a series of vibrations on the earth's surface caused by the generation of elastic (seismic) waves during the sudden rupture and release of accumulated strain energy due to the deformation caused by the constant movement of plates. Earthquakes are the result of plate tectonics or shifting plates in the earths crust and quakes occur when the frictional stress of the gliding plate boundaries builds and causes failure at a fault line.

Severity of an earthquake is measured on a magnitude scale known as Richter Scale and on the basis of this scale, the earthquakes are classified as:

- Slight (Magnitude up to 4.9 on Richter Scale)
- Moderate (Magnitude up to 5–6.9 on Richter Scale)
- Great (Magnitude up to 7.0–7.9 on Richter Scale)
- Very Great (Magnitude up to 8.0 and more)

A proactive stance to reduce the toll of disasters in a region requires a more comprehensive approach that encompasses both pre-disaster risk reduction and post-disaster recovery (Kramer 1996). It is framed by new policies and institutional arrangements that support effective action. Following set of activities are needed for the comprehensive approach of earth quick management (NDMA 2006):

- Risk analysis to identify the kind of risk faced by the people
- Prevention and mitigation, to address the structural sources of vulnerability
- Emergency preparedness and response, to enhance a country's readiness to cope quickly and effectively with the problems of a disaster
- Post-disaster rehabilitation and reconstruction, to support effective recovery and to safeguard against any future disasters

Before adopting a control measure, the true nature of an earthquake must be understood properly and for this different theories and models were put forward from time to time about the causes of an earthquake (Lee et al. 1987).

Worthy to note is the Dilatancy—Diffusion Theory developed in the United States and the Dilatancy—Instability theory developed in the then USSR.

The interesting fact is that the first stage of both the models is an increase of elastic strain in a rock that causes them to undergo a dilatancy state, which is an inelastic increase in volume that starts after the stress on a rock reaches one-half its breaking strength. Hence it is in this state the first physical change takes place indicating any future earthquakes.

The USA model suggests that the dilatancy and fracture of the rocks is first associated with low water containing dilated rocks, which helps in producing lower seismic events. The pore water pressure then increases due to influx of water into the open fracture, weakening the rock and facilitating movement along the fracture, which is termed as an earthquake (NDMA 2006).

The Russian Model suggests that the first phase is accompanied by an avalanche of fractures that releases some stress but produces an unstable situation that eventually causes a large movement along a fracture.

Seismic gaps, the areas along active fault zones, capable of producing large earthquakes have not recently produced any earthquake. It is these areas which are thought to bring in tectonic strain and are the candidates for future large earthquakes. Any fault that has moved during quaternary can be called as active fault and it is generally assumed that these faults could get displaced at any time (Babar 2007).

The active faults are basically responsible for seismic shaking and surface rupture. Faults that have been inactive for the last 3 million years are generally classified as inactive fault. Like all other natural hazards earthquakes also produce primary and secondary effects. Primary effects include surface vibration, which may be associated with surface rupture and displacement along fault planes. These vibrations may sometimes lead to the total collapse of large buildings, dams, tunnels, pipelines, and other rigid structures (NDMA 2006).

Secondary effects of an earthquake include a variety of short range events (such as liquefaction landslides, fires, tsunamis, and floods) and long-range effects including regional phenomenon (such as regional subsidence or emergence of landmasses, river shifting and regional changes in ground level).

The main objective of earthquake preventive measures should be to develop and promote knowledge, practices and policies that reduce fatalities, injuries, and other economic losses from an earthquake. Geographic Information System and Remote Sensing, provide an effective and efficient tool for the storage and manipulation of the remotely sensed data and other spatial and nonspatial data types for both scientific management and policy-oriented information. This can be used to facilitate measurement, mapping, monitoring, and modeling of variety of data types related to such natural phenomenon.

Effective Earthquake Management needs a special focus on the following critical areas:

- Awareness about the seismic risks
- Integration of the structural mitigation measures in engineering syllabus
- Monitoring and enforcement of earthquake-resistant building codes and town planning by laws
- Licensing of engineers and masons
- Earthquake-resistant features in non-engineered construction in suburban and rural areas
- Training among professionals in earthquake-resistant construction practices
- Adequate preparedness and response capacity among various stakeholder groups

Following are a few personal dos and don'ts that should be followed at the time of an earthquake.

Do's
- Take shelter under a table, desk, bed, or doorway during an earthquake.
- Provide help to others and develop confidence.
- Shut off the kitchen gas.
- Keep stock of food stuff, drinking water, and first aid arrangements.
- If you are in a moving vehicle, stop and stay in the vehicle.
- Follow and advocate local safety building code for earthquake-resistant construction.
- Keep the heavy objects and glasses on lower shelves.
- Keep yourself updated by the latest information.

- Make plans and preparations for emergency relief.

Don'ts
- Do not get panicky.
- Do not use candles, matches etc. and do not switch any electric mains immediately after an earthquake.
- Do not spread and believe in rumors.
- Do not run through or near buildings during an earthquake.

Disposal of Hazardous Wastes

Production of wastes has been an outcome of the human societies since prehistoric times and the same will continue to happen in the future as well. So the disposal of wastes into the surrounding locality has been a usual practice with little concern for the environment. Wastes need to be managed properly to preserve the planet for the coming generations. Rapid trend of industry and high-technological progress are the main sources of the accumulation of hazardous materials. Nuclear applications have developed at a rapid rate and several nuclear power plants have been started to work throughout the world (Bhoyar et al. 1996). The potential impact of released radioactive contaminants into the environment has received growing attention due to nuclear accidents, which pose serious problems to biological systems.

Wastes are classified as hazardous, if they exhibit one or more of ignitability, corrosivity, reactivity, or toxicity. According to Resource Conservation and Recovery Act (RCRA), hazardous wastes are defined as any waste or combination of wastes which pose a substantial or potential hazard to human health or living organisms because such wastes are nondegradable or persistent in nature or because they can be biologically magnified, or because they can be lethal, or because they may otherwise cause or tend to cause detrimental cumulative effects. Hazardous wastes refer to wastes that may, or tend to, cause adverse health effects on the ecosystems and human beings. These wastes pose present or potential risks to human health or living organisms, due to the fact that they are nondegradable or persistent in nature and can be biologically magnified. These wastes are highly toxic and even lethal at very low concentrations. Hazardous wastes contain toxic substances generated from industrial, hospital, some types of household wastes. These wastes could be corrosive, inflammable, explosive, or react when exposed to other materials. Some hazardous wastes are highly toxic to environment including humans, animals, and plants. With increasing manufacturing processes, solid, liquid, and/or gaseous emissions generate as by-products. Some of these wastes are potentially harmful to human health and environment and thus need special techniques of management. The management of hazardous wastes has become a specialized discipline because of the complex nature of the problem and the solutions available to humanity. The mismanagement examples of hazardous wastes causing disastrous human and environmental consequences are numerous. The management process is based on the definition and classification of the different wastes, and their toxic effects on human and taking in consideration the application of risk management to control human health and environmental impacts of hazardous waste. Hazardous waste management, therefore, deals with minimizing harmful effects on humans and environment by applying special techniques of handling, storage, transportation, treatment, and disposal of hazardous wastes.

(a) Industrial wastewater
(b) Radioactive wastes
(c) Biomedical wastes
(d) Chemical wastes

Industrial Wastewater Disposal Wastewater from industries include cooling water, stormwater, processed water, and washed down waters (Ahuja and Abda 2015), is characterized by high pollutant loads that may include colloidal, suspended, and dissolved (mineral and organic) solids. In addition, industrial wastewater may also include some inert, organic, or toxic materials and pathogenic microbes (Ahuja and Abda 2015). Composition of industrial wastewater varies depending on their origin and treatment technology applied. For instance, wastewater obtained from agro and food industries is rich in biodegradable substances, which can result in the depletion of dissolved oxygen in the aquatic environment, hence posing a serious threat if discharged without proper treatment. Industrial wastewater thus poses a potential threat to the environment and human health, demanding wastes to be removed from production sources and treated suitably before disposal. So, the objective in the adequate management and appropriate treatment of industrial wastewater is the achievement of the environmental standards and solution of some other related concerns.

Treatment of Industrial Wastewater Based on the type of industrial wastewater, the specific treatment process that requires both the comprehensive understanding of the production process and the system organization is employed. There are four types of industrial wastewater/effluents that need to be considered:

1. General manufacturing effluents: These are the effluents resulting from the mixture of water with solids, liquids, or gases, which may include intermittent effluents like the ones produced from agri–food industry or the continuous effluents like the ones produced from pharmaceutical and parachemical industries. While the former usually have regular effluent flow the latter one includes effluents that are more difficult to analyze as they keep on changing.
2. Specific effluents: Some effluents, such as those obtained from pickling and electroplating baths, need to be separated either for specific treatment after which they are recovered or they may be stored in tanks before being reinjected at a weighted flow rate into the treatment line.
3. General service effluents: These effluents may include wastewater from boiler blowdown, spent resin regenerants, etc.

4. Intermittent effluents: These may include wastes resulting from accidental leaks during handling or storage of products and may also include wastewater from floor wash, storms, etc.

Since the industrial wastewater possesses different types of contaminations it requires a variety of strategies to treat it. Some of these are:

1. **Removal of Solids:** Simple sedimentation techniques are employed for the removal of solids from the wastewater. The removed solids are recovered as sludge or slurry. However, for very fine particles or solids with densities similar to the density of water, filtration, or ultrafiltration is used.

2. **Removal of Oil and Grease:** To achieve the water purity of desired level by removing the oils from wastewater, the oil skimmers are considered as cheap and dependable instruments. Skimming is preferred before using membrane filters and chemical processes as it prevents the premature blinding of the filters. However, for greases, the skimmers are to be fitted with heaters to keep the greases in fluid form for its proper discharge, as the greases contain hydrocarbons of higher viscosity.

3. **Removal of Biodegradable Organics:** For the removal of biodegradable organics, conventional wastewater treatment processes like trickling filters and activated sludge methods are used. However, if the wastewater is heavily loaded with cleaning agents, pesticides, disinfectants, or antibodies, it can have an adverse impact on the treatment process.

4. **Removal of Toxic Materials:** They include acids, heavy metals, alkalis, nonmetallic elements resistant to biological degradation. They are precipitated out by bringing about a change in the pH of the solution or by treating the materials with other chemicals. However, wastes resistant to these treatments are concentrated prior to recycling or landfilling. Further, dissolved organics present in the wastewater are incinerated within the wastewater using different techniques such as advanced oxidation processes.

5. **Removal of Other Organics:** Pesticides, pharmaceutical products, solvents, paints, cooking products, etc., are difficult to treat. Hence, different methods such as distillation, incineration, adsorption, chemical immobilization, vitrification, or advanced oxidation processes are used for their treatment.

Steps included in the disposal of industrial wastewater include: (a) Reducing the quantity of the wastewater, (b) Recycling the wastewater for flushing, (c) Reusing the wastewater (after treatment) for a beneficial purpose (either on a neighboring property or onsite) such as for growing crops, in gardens or in turfs, (d) Treating and discharging the wastewater to sewer, (e) Treating and discharging the wastewater to lined or soakage evaporation pit, (f) Treating and discharging the wastewater to watercourses or drains. After proper treatment of the wastewater, it has to be disposed of by reintroducing into the environment. Three commonly used methods for disposing the treated wastewater effluents are: **(a) Discharge into Surface Water:** Also referred to as disposal by dilution, in this method the industrial wastewater after proper treatment is either reused or released into the sanitary sewers or into the surface waters. This is generally used where the volume, degree, and composition of the wastewater is less in comparison to receiving water. **(b) Surface Disposal:** Also referred to as disposal by irrigation, it involves spreading of the wastewater over the surface by constructing irrigation channels or ditches. With little evaporation, a major portion of the wastewater soaks into the ground, thus becoming a source of moisture and nutrients to plants growing in that area (LaGrega et al. 1994). However, this method is restricted to moderately little volumes of wastewater from a generally small population where there is the availability of land and where nuisance problem is less likely to occur. This method has its wide application in arid or semiarid areas where the scarcity of moisture in the soil can be checked by the surface disposal. **(c) Subsurface Disposal:** This method involves the introduction of wastewater into the underground surface through pits or tile fields. Subsurface disposal is commonly used for the disposal of settled industrial wastewater with limited volume of discharge thus lacks wider application on a larger scale.

Disposal of Radioactive Wastes

Radioactive substances present naturally in the environment are used in a number of ways by mankind such as medical diagnosis, therapy, generation of electricity, scientific research, and some other specialized industrial applications (IAEA 1995). However, these operations, using radioactive materials, generate wastes such as gases, liquids, or solids that need to be managed appropriately subjected to the standards of safety. After proper treatment the air-borne and liquid waste can be discharged into the environment; however, the solid wastes can be disposed-off only at appropriate sites or stored until a suitable disposal route becomes available (Freiesleben 2013). Based on the level of radioactivity, these wastes are classified as exempt, low-level, intermediate, and high-level wastes. Additionally, the solid waste is further divided into the primary waste that may include machinery and its parts contaminated with radioactivity substances (metallic hardware) and secondary waste that may include the waste produced during various operational actions. It is necessary to characterize the waste precisely as its quantity directly affects the treatment process to be chosen. For the safety of human beings and protection of the environment, these radioactive wastes need to be managed in a proper way. In addition to this, the safety of waste management facilities has to be appropriately assured during their lifetime. One of the important steps in the management of radioactive waste is to confine and segregate the waste from the biosphere, for which it is generally processed to convert it into solid and stable forms. Further, the total volume of the waste is reduced and it is immobilized in order to facilitate its storage, transport, and disposal (Freiesleben 2013).

For the disposal of radioactive wastes, the following are different available procedures:

1. **Specific Landfill Disposal:** Similar to conventional landfill disposal, it is designed for low-level radioactive waste, possessing low concentration or quantity of radioactive substances.
2. **Near Surface Disposal:** It includes construction of engineered vaults or trenches on the surface or few meters below the surface. This process is again used for the disposal of low-level radioactive waste.
3. **Disposal of Intermediate Level Waste:** It involves the construction of vaults, caverns, or silos at least tens to hundreds of meters below the surface level. This type of disposal facility may include existing mines or those developed by drift mining (in which overlying cover is more than 100-m deep).
4. **Geological Disposal:** Here the waste is disposed in vaults, tunnels, or silos that are constructed in particular geological formations possessing long-term stability and located at least a few hundred meters below the ground level. Geological disposal is designed to receive high-level radioactive waste.
5. **Borehole Disposal:** It includes a single borehole or an array of boreholes between tens of meters up to a few hundred meters deep. Such facilities are designed for relatively smaller quantities of radioactive waste. Although it can be used for high-level solid waste disposal and spent fuel but has not been adopted so far.
6. **Disposal of Mining and Mineral Processing Waste:** It involves the disposal of waste on or near the ground after being stabilized in situ and then covered with several layers of soils and rocks.

The selection of a particular method or disposal facility depends on various factors including progressively improving data, technical and regulatory reviews, political decisions, and public consultation.

Disposal of Biomedical Wastes

These are some hazardous medical and dental wastes which, when disposed improperly, could hurt the earth or these are the human services waste which are equipped to deliver damage or illness. Numerous sorts of hazardous biological wastes include:

Infectious waste: Wastes which that contain pathogens like viruses, bacteria, fungal spores, or other parasites in concentrations adequate enough to cause illness in a host are known as infectious wastes. Such wastes like dressings and tissues are usually generated from research laboratories, diagnostic laboratories, postmortem examinations, medical procedures, and treatment of infected patients and creatures or are the materials or gear in contact with blood and contaminated body fluids (MoEF 1998).

Pathological waste: Wastes including body parts, tissues, organs, human fetuses, animal carcasses, blood and other body fluids are known as pathological wastes.

Sharps: Sharps contain syringes, needles, surgical tools, saws, blades, broken glass, or different things that cause cuts or puncture wounds.

Pharmaceutical waste: It covers terminated, unused, spilt, and sullied pharmaceutical items, medications, antibodies, and sera that are never again required and should be disposed of in a suitable way.

Genotoxic waste: This category includes consolidates cytostatic drugs, regurgitation, pee, or defecation from the patients treated with cytotoxic medications, synthetic compounds, and radioactive materials. Genotoxic waste has mutagenic, teratogenic, and cancer-causing properties (MoEF 1998).

Chemical waste: Discarded strong fluid, or vaporous synthetic compounds ought to be considered as hazardous on the off chance that it is lethal, destructive, inflammable, or responsive.

Waste with high substance of substantial metals: Mercury (thermometers, circulatory strain checks, amalgam), cadmium (disposed of batteries), and lead (strengthened wood boards for radiation sealing in radiology offices) produced from healing centers could be taken to as a subcategory of hazardous synthetic wastes.

Stages of Medical Waste Disposal

Stage 1: Collection and Segregation
During the handling process, the biomedical waste must be collected in strong and resilient containers, with no sharps, syringes, needles, or other contaminated items placed in the common waste or recycle bins, as this causes all waste to become infectious. Separation of the liquid and solid biomedical waste products shall be carried out. Categorization of the medical waste with proper segregation in color-coded containers, in order to isolate and properly manage each waste, shall be done.

For the purpose of isolation, biomedical waste is generally classified into following eight catagories:

- General: Printed materials, sustenance wastes, bundling materials, and so on, usually non risky to human health as no blood or any other body fluid is attached to these wastes.
- Radioactive: Unused radiotherapy or laboratory research liquids and other contaminated glassware etc.
- Pharmaceutical: unused, expired and contaminated medicines.
- Sharps: Wastes like needles, blades, knives, scalpels, scissors and other sharp wastes.
- Pathological: Tissues or body fluids like body parts, blood, pus and other body fluids.
- Infectious: Waste that can pass on viral, bacterial, and parasitic infections to humans, i.e. laboratory cultures, tissues, swabs, equipments and excretory products.
- Chemical: Cleaning agents, disinfectants, expired laboratory reagents, film developers, other chemical substances of use.

- Pressurized containers: Barrels containing pressurized gas, gas cylinders and gas cartridges etc.

Stage 2: Storage and Transportation

Specific storage requirements, such as a secure area that is inaccessible to the public and which separates it from food consumption areas should be available for the proper storage of biomedical wastes. The storage units must also be accompanied by a fridge or freezer unit which, if needed, can be used for the medical waste. Special vehicles and protective devices should also be available for the proper disposal, manipulation, or transport of the biomedical waste products. Further, the protective devices should be regularly observed and maintained so that they do not become the sources of infection transmission (Altaf and Mujeeb 2002).

Stage 3: Treatment

It requires the professional handling of regulated medical waste (RMW) in accordance with the statutory regulations, such as OSHA (Occupational Safety and Health Association). In the process, several medical waste equipment such as handling carts, scrapping, transport, reduction of size, compactors, sterilization, or recycling, are used. In order to reduce the risks and preserve the environment, the following equipment are required to properly process the waste (MoEF 1998):

- Carts and containers—generally used for medical waste collection, e.g., dumpers, containers, and compactor
- Conveyors—these devices help to separate the waste
- Sterilizers—e.g., autoclaves used to sterilize the waste
- Waste handling—e.g., compactors, containers, pre-crushers, and deliquefying systems
- Recycling System—e.g., the equipment of size reduction

Incineration System In the combustion process, the incineration technology uses a high-temperature heat process that converts inert materials and gases into ash, gas, and heat. Three types of incinerators are commonly used for the treatment of biomedical waste: (1) **The Multiple Hearth Type:** It has a circular stainless steel furnace with solid refractory fireplaces and a central rotating shaft, which converts the waste into ash. (2) **Rotary Kiln:** It is a drum-like incinerator commonly used for medical and hazardous waste. (3) **Controlled Air:** There are two process chambers that handle the waste and are usually used in waste containing organic materials. The total combustion and oxidation lead to a stream of gas with a composition of carbon dioxide and water vapors.

Non-incineration System The non-incineration techniques provide for the disposal of biomedical waste through irradiative, thermal, chemical, and biological methods. The thermal system requires high (thermal) temperatures that produce steam for the removal of biomedical waste, as the steam plays a crucial role in the process of autoclaving the waste in a container that holds it. Although this technique can be used for all types of human body fluids, autoclaving cannot degrade the cytotoxic agents used for chemotherapy. Irradiation is another non-incineration technique that treats the waste using high-frequency microwaves. The microwaves produce heat that kills bacteria or other contaminants. Chemical decontamination is another way to treat the biomedical waste, which can be used in the case of microbiology laboratory residues, human blood and other body fluids, and sharp wastes, but cannot be used for anatomical wastes. Further, the method of biological processes can also be used to treat the organic wastes by the application of specific enzymes.

Onsite and Offsite Treatment of Medical Waste Whether to use the onsite or offsite mode of treatment, it is important to distinguish between the biomedical waste, because the biomedical waste in most of the instances is present in a mixed form. It often becomes very difficult to manage or even to separate the waste. The onsite treatment usually entails some costly equipment and a major infrastructure expenditure, which is generally cost-effective for very big hospitals and laboratories. Most medical waste producers, therefore, opt for an offsite treatment known as regulated companies of medical waste disposal because they have the following requirements for the same: (a) State certified operating permits, (b) OSHA trained personnel to collect, transport and store the medical waste, (c) Availability of proper medical waste equipment.

Stage 4 Disposal: In the United States, for example, they can use their municipal and sanitary landfills and sewer systems as a final disposal method, once medical waste producers have complied with the rules for collection, storage, transport, and treatment of their waste. So, the municipal site is often used as the final site for the treated biomedical wastes. Almost every state and local government has its own rules and regulations for the fluids, such as blood, suction fluids, excretions, and secretions, and the two generally recommended ways of dealing with medical waste fluids are: (a) Collection of fluids in a leak-proof autoclavable container, (b) Disposal of thermally autoclaved fluids is the sanitary sewerage systems. However, before discharging the treated fluids into the sewers, additional precautions should be taken because they can obstruct and can leak the system.

Disposal of Chemical Wastes

Waste from industries that possess potential danger to human health or the environment because of its quantity, concentration, and properties (physical and chemical) is known as chemical waste (Bill 2010). It includes toxic pollutants, acids, bases, organic, and inorganic matter high in biological oxygen demand and suspended solids in varying degrees of concentrations. Many materials in the chemical industry are toxic, carcinogenic, mutagenic, or nonbiodegradable hence need to be treated at the source. In chemical industries, the extensively used emulsifiers, surfactants, and petroleum hydrocarbons usually result in a decreased efficiency of treatment plants. Generally, the chemical waste from the industries varies from the domestic sewage in composition, hence pretreatment is required to produce an equivalent effluent.

Further, the biodegradability, molecular size and high variability of the pollutants, stringent effluent permits, and extreme operating conditions define the practice of wastewater treatment. First and foremost step in the treatment process involves the minimization of waste during the production. Some common treatment methods for chemical industrial wastewater include biological oxidation methods such as rotating biological contactors, trickling filters, activated sludge, or lagoons. Coagulation methods followed by sedimentation or flotation is also used to treat the large molecular size pollutants. However, the waste that cannot be reused or recycled needs to be disposed-off properly and for the set-up of the disposal facilities, guidelines framed by the Pollution Control Boards needs to be followed strictly. The procedures for the management/disposal of chemical waste are as follows:

1. **Waste Minimization:** Prior to disposal, reduction in volume, and toxicity of waste is carried out to reduce the potential health hazard, pollution, and costs for various parties, thereby reducing potential long-term liabilities for disposal. It can be achieved by limiting the amount of chemicals that are purchased, substitution of raw material with less hazardous chemicals, segregation of waste streams, clear labeling of containers, neutralizing, quenching, or otherwise destroying hazardous by-products in the final step of an experimental procedure.

2. **Waste Segregation:** Segregation of hazardous and nonhazardous waste is necessary for the proper disposal of chemical waste. Since different wastes are disposed by various methods, they are also reflective in the cost of disposal. For instance, nonhalogenated organics are less expensive to dispose-off than halogenated organic wastes that are destroyed in a chemical incinerator. Further accidental mixing of hazardous and nonhazardous chemicals can lead to fire, explosion, and other physical and chemical hazards. In addition, the maintenance of safety data sheets that lists the incompatible chemicals can aid in determining which waste stream is hazardous and how to deal with it.

3. **Storage:** Before disposing, the chemical waste must be packed without the danger of leakage, explosion, or escape of hazardous vapors, that allows it to be stored and transported. Containers used to store these routine chemical wastes are filled to only 80% of the total volume capacity to allow for vapor expansion and to minimize the spills that could occur during the movement of overfilled containers. Further, these containers should be stored away from direct sunlight and ignition sources.

4. **Disposal:** The method that could be used for disposal of waste depends upon the degree of treatment before disposal, selection of treatment technology, and onsite/offsite methods. Methods thus used for the disposal of hazardous chemical waste include **(a) Land Disposal:** In this method, the disposal of waste is carried out by mixing it with surface soil thereby transforming, degrading or immobilizing the waste through proper management. This type of disposal is suitable for hazardous oil sludges from petroleum refineries and other biodegradable wastes. However, flammable, reactive, highly volatile, liquids and inorganic wastes including acids, bases, and heavy metals are not disposed using this method. **(b) Deep Well Injection:** For those wastes that cannot be disposed using land disposal, deep

well injection, in which liquid waste is pumped through a steel coating into a porous layer of limestone or sandstone is used. Liquid wastes are stored into the pores and fissures of the rocks by applying high pressure and the waste is stored there for a longer period of time. The injection zone must lie below the impervious rock layer and should extend more than 0.8-km below the surface. One of the advantages of deep well injection is that it requires little or no pre-treatment of the waste. **(c) Incineration:** It involves the combustion of wastes to convert them into base components with the released heat being trapped for deriving the energy. Here, the volume of waste is reduced by about 90% with assorted gases and inert ash as a common by-product. **(d) Ocean Dumping:** Wastes that are highly toxic in nature are generally dumped in oceans far from human habitation. It involves the dumping of wastes such as industrial waste, sewage sludge, dredged material, and radioactive wastes.

Disposal of Empty Containers and Glassware

- Remove the barcode label (if applicable) and return it to the Dispensing Chemist so the chemical can be removed from the Chemical Inventory.
- Rinse the container several times with water to remove all residues.
- If the contents of the container are not water soluble, rinse the container several times with an appropriate solvent.
- Rinsate should be collected and disposed of as hazardous liquid chemical waste if the container previously held a hazardous chemical or was cleaned using a solvent.
- If necessary, allow the residue to evaporate in a fume hood before the container is discarded. Containers rinsed with water may be directly disposed off.
- Remove or deface the container label with an X.
- Discard glass containers in the white, plastic buckets provided for glass waste. Other containers may be disposed of in the trash.

References

Ahuja, A. S., & Abda, S. A. (2015). Industrial hazardous waste management by government of Gujarat. *Research Hub International Multidisciplinary Research Journal, 2,* 1–11.

Alaric, S. V., & Patrick, B. R. (eds) (2014). Forest conservation and management in the anthropocene: Conference proceedings. Proceedings. RMRS-P-71. Fort Collins, CO: US Department of Agriculture, Forest Service. Rocky Mountain Research Station. 494 p.

Altaf, A., & Mujeeb, S. A. (2002). Unsafe disposal of medical waste: A threat to the community and environment. *Journal of Pakistan Medical Association, 56*(2), 232–233.

Arnalds, A. (2005). Approaches to landcare – A century of soil conservation in Iceland. *Land Degradation and Development, 16,* 113–125.

Babar, M. (2007). Environmental changes and natural disasters. Pitam pora New Delhi: New India Publishing Agency. 229 pp.

Bhoyar, R. V., Titus, S. K., Bhide, A. D., & Khanna, P. (1996). Municipal and industrial solid waste management in India. *Journal of IAEM, 23,* 53–64.

Bill, H. (2010). Techniques for efficient hazardous chemicals handling and disposal. *Pollution Equipment News*, p. 13. Retrieved March 10, 2016.

Chapin, F. S., Matson, P. A., & Mooney, H. A. (2011). *Principles of terrestrial ecosystem ecology* (2nd ed.). New York: Springer. isbn:0-387-95443-0.

Clerici, A., Perego, S., Tellini, C., & Vescovi, P. (2002). A procedure for landslide susceptibility zonation by the conditional analysis method. *Geomorphology, 48*(4), 349–364.

Ellis-Jones, J., & Sims, B. (1995). An appraisal of soil conservation technologies on hillside farms in Honduras, Mexico and Nicaragua. *Project Appraisal, 10*, 125–134.

Freiesleben, H. (2013). Final disposal of radioactive waste. *EPJ Web of Conference*. https://doi.org/10.1051/epjconf/20135401006.

Govers, G., Merckx, R., Wesemae, B. V., & Oost, K. V. (2017). Soil conservation in the 21st century: Why we need smart agricultural intensification. *The Soil, 3*, 45–59.

Groom, M. J., Meffe, G. K., & Carroll, C. R. (2006). *Principles of conservation biology* (3rd ed.). Sunderland, MA: Sinauer Associates.

Hunter, M. L. (1996). *Fundamentals of conservation biology*. Cambridge, MA: Blackwell Science.

International Atomic Energy Agency, IAEA. (1995). *The principles of radioactive waste management*, IAEA Safety Series No. 111-F.

Kramer, S. L. (1996). *Geotechnical earthquake engineering*. Upper Saddle River, NJ: Prentice Hall.

Kumar, V. (2015). Social forestry in India: Concept and schemes. *Van Sangyan, 2*(11), 18–22.

LaGrega, M. D., Buckingham, P. L., & Evans, J. C. (1994). *Hazardous waste management*. New York: McGraw-Hill. 1146 pp.

Lee, S. L., Balendra, T., & Tan, T. S. (1987). A study of earthquake acceleration response spectra at far field. In *US-Asia conference on engineering for mitigating natural hazards damage*. Bangkok, Thailand, 14–18 December 1987.

Loch, R. J. (2004). *Soil conservation practice – in search of effective solutions*. In ISCO 2004 - 13th International Soil Conservation Organisation Conference – Brisbane, July 2004. Conserving Soil and Water for Society: Sharing Solutions (pp. 1–6).

MoEF. (1998). *GoI, The gazette of India: Extraordinary, notification on the bio-medical waste (Management and Handling) rules, [Part II—Sec.3(ii)], No. 460* (pp. 1–20). New Delhi: Ministry of Environment and Forests.

National Disaster Management Authority (NDMA). (2006) *Draft National Disaster Management Framework*, pp. 14–17.

National Disaster Management Guidelines (NDMG). (2009, June). *Management of landslides and snow avalanches*. New Delhi: National Disaster Management Authority, Government of India.

Panos, P., Pasquale, B., Katrin, M., Christine, A., Emanuele, L., & Luca, M. (2015). Estimating the soil erosion cover-management factor at the European scale. *Land Use Policy, 48*, 38–50.

Pawar, K. V., & Rothkar, R. V. (2015). Forest conservation and environmental awareness. *Procedia Earth and Planetary Science, 11*, 212–215.

Prasad, B. C. (1985). Voluntary Agencies: A Movement to Conserve Forest Wealth. *Indian and Foreign Review 22*(7):9–10, 29.

Pretty, J., & Shah, P. (1994). *Soil and water conservation in the twentieth century: A history of coercion and control*. Research Series No 1. Reading: Rural History Centre, University of Reading.

Rapport, D. J., Costanza, R., & McMichael, A. J. (1998). Assessing ecosystem health. *TREE, 13*, 397–402.

Ros-Tonen, M., Zaal, F., & Dietz, T. (2005). Reconciling conservation and livelihood needs: New forest management perspectives in the 21st century. In *African forests between nature and livelihood resources* (p. 8). Queenstown: Edwin Mellen Press.

Sayer, J. A., & Maginnis, S. (2005). *Forests in landscapes: Ecosystem approaches to sustainability*. London: IUCN/Earthscan.

Singh, A. K. (2010). Landslide management: Concept and philosophy. *Disaster Prevention and Management: An International Journal, 19*(1), 119–134.

Soule, M. E., & Wilcox, B. A. (1980). *Conservation biology: An evolutionary-ecological perspective*. Sunderland, MA: Sinauer Associates.

Williams, B. K. (2011). Adaptive management of natural resources: Framework and issues. *Journal of Environmental Management, 92*, 1346–1353.